CANAL IRRIGATION IN
PREHISTORIC MEXICO

Canal Irrigation in Prehistoric Mexico

The Sequence of Technological Change

William E. Doolittle

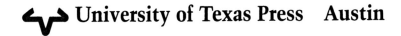 University of Texas Press   Austin

Drawing of irrigator on title page by John V. Cotter. After the Florentine Codex.

Copyright © 1990 by the University of Texas Press
All rights reserved
Printed in the United States of America

First paperback edition, 2011

Requests for permission to reproduce material from this work should be sent to Permissions, University of Texas Press, Box 7819, Austin, TX 78713–7819.

⊗The paper used in this publication meets the minimum requirements of American National Standard for Information Sciences—Permanence of Paper for Printed Library Materials, ANSI Z39.48–1984.

Library of Congress Cataloging-in-Publication Data

Doolittle, William Emery.
    Canal irrigation in prehistoric Mexico : the sequence of technological change / William E. Doolittle. — 1st ed.
        p.   cm.
    Bibliography: p.
    Includes index.
    ISBN: 978-0-292-72953-7
    1. Indians of Mexico—Irrigation. 2. Canals—Mexico. 3. Irrigation—Mexico. 4. Indians of Mexico—Antiquities. 5. Mexico—Antiquities.
I. Title.
F1219.3.I77D66   1990
627'.52'09720902—dc20   89-32177
                                           CIP

Once again,
for all the men and women
who died or remain missing
in Southeast Asia

Contents

Figures

Tables

Preface

Gods of water comprised a great deal of the religious life of ancient Mexican societies dependent on agriculture. To the Aztecs, Tlaloc, "he who makes things grow," was the most important hydraulic deity. Typically portrayed carrying a vase of water, he was generally a propitious god. Tlaloc was, however, also a god whose anger was much feared. Priests commonly sacrificed prisoners, especially children dressed like him, in order to make appeasements or to have requests of water granted (Caso 1937 : 27–28).

In spite of their beliefs and fears, prehistoric farmers in Mexico not only attempted but succeeded in creating a means of freeing themselves, at least in part, from the capriciousness of deities such as Tlaloc. Specifically, they invented irrigation, developed it into a science, and used it widely.

Irrigation has many forms and can involve a number of different practices (Rojas Rabiela 1985 : 192–210). Watering plants by means of hand-held pots, redirecting the flow of runoff from slopes to fields, and planting in pits excavated in order to bring roots in contact with subsurface water are only a few examples of irrigation techniques. Most commonly, however, irrigation is equated with networks of canals through which water is transported from a source to fields that would otherwise be dry. Known specifically as canal irrigation, this particular agricultural practice has garnered a great deal of attention from prehistorians, probably because remnants of ancient canals are abundant and, in some locales, sufficiently extensive to indicate that this form of hydraulic manipulation played a significant role in the rise of high cultures.

To date there have been numerous investigations of individual relict canal irrigation systems. Lacking, however, is a study that synthesizes the data collected during these previous research activities in order to gain a broader understanding of this agricultural practice as a form of environmental modification. Similar studies have been conducted on other types of ancient agriculture. For example, Robin Donkin (1979) synthesized data collected by others for his study of prehistoric terracing on slope lands, and J. P. Darch (1983) did likewise in her volume on raised fields built in order to cultivate wet lands. The present study follows in this tradition. It is the first analysis of prehistoric cultivation of dry lands by means of canal irrigation that brings together data collected in other studies.

As in the case of those studies that have synthesized data on terracing and raised fields, this one has a regional orientation. Here, however, the region is comparatively small. Whereas Donkin's book included the entire New World, and Darch's volume dealt with Central and South America, this one is limited to Mexico, and more specifically only two-thirds of that country. This region is the semi-arid northern half of Mesoamerica and its arid northern frontier area. Excluded are all other areas where canal irrigation was practiced in prehistoric times, including the American Southwest.

A good argument could be made for the inclusion of the Southwest. Certainly, canal irrigation was practiced there prehistorically, and it may well have been relatively more important than it was in Mexico. The reason why the Southwest is excluded is really quite simple—sufficient information on prehistoric canal irrigation exists to merit two volumes, one on Mexico and one on the American Southwest.

The research on which this book is based began as a study of ancient canal irrigation in the Southwest. The search for the origins of canal irrigation there quickly led to an investigation of data from Mexico. The known existence of irrigation canals in Mexico, predating those in the Southwest, and the long-recognized cultural connections between the two areas required that at least a cursory assessment of early Mexican irrigation be undertaken. Approximately halfway through that phase of the project, however, it became apparent that canal irrigation in the two areas developed along completely different lines (this is discussed in parts of Chapters 3 and 4). It was decided, therefore, to conduct a separate and in-depth study of ancient canals in Mexico first, and to complete the study of prehistoric canal irrigation in the American Southwest later. That study is once again under way.

In addition to being different in terms of the scale of its regional orientation, this study is unlike other studies that have synthesized data on specific prehistoric agricultural practices in that it does not focus on fields or planting surfaces. As in the case with both terraces and raised fields, canals and their related features can be considered as "agricultural landforms," modifications of surface relief related to cultivation (Golomb and Eder 1964; Nir 1983 : 37–62). In this study, however, they are treated as technology.

Charles Singer, E. J. Holmyard, and A. R. Hall (1954–1958 : 1 : vii), in their five-volume *History of Technology*, define technology rather loosely as "how things are commonly done or made . . . and what things are done and made." In the French counterpart of this monumental British work, the *Histoire Générale des Techniques* edited by Maurice Daumas (1962–1968), technology is also considered to be synonymous with both techniques and things produced by techniques. Although both of these works are considered "classics" by most scholars, the validity of their definition has been questioned by some, especially American scholars, because it does not emphasize the importance of the "improver of technology" (Multhauf 1959) and the "knowledge" possessed and applied (Layton 1974).

These criticisms hold especially true for relatively recent technology, particularly that of the so-called "Western World," where scientific leadership is important in intellectual development (White 1966 : 99–100; Kranzberg and Pursell 1967 : 1 : 6). Indeed modern societies employ groups of engineers who systematically plan technological improvements. For earlier, and especially ancient societies, however, the vague, imprecise, and general definition of technology proffered by European scholars seems to be more appropriate given the nature of archaeological data. The "things" that were "done and made" long ago certainly can be identified archaeologically. "How" they were "done and made" can be inferred with reasonable accuracy. The degree to which early irrigators actually thought of themselves as "improvers of technology," however, will never be known. Similarly, we can only speculate as to how these people gained "knowledge."

The justification for emphasizing technology and its development is outlined in Chapter 1. Suffice it to say for now that the emphasis on technology is intended to fill a void in our understanding of prehistory, especially that of the Americas, and particularly our understanding of how ancient cultures sustained themselves. Most studies of prehistoric agriculture are undertaken by archaeologists trained as anthropologists. As a result, they tend to focus on the organization of irrigation societies, particularly their political, economic, and cultural structures. *Mesoamerica: The Evolution of a Civilization* by William T. Sanders and Barbara J. Price (1968) is perhaps the best example of studies of this type. Such studies are often of a cultural ecological nature, involving an attempt to understand how cultures "adapted" to their bio-physical environment (Kirch 1980). Although they typically discuss in a general sense how people modified the environment to fit their needs, they do not elaborate on the *mechanics* involved (see Ford 1977 : 139−141). Technology is certainly a critical factor in the human-environmental relationship (Carlstein 1982 : 5−9). Given its inherently geographical nature, it is appropriate for someone trained in that discipline to address its development.

This work was completed only with the help of my home institution and numerous people. Funding was provided by a Mellon Faculty Summer Research Grant through the Institute of Latin American Studies, and two Research Grants from the University Research Institute, all of the University of Texas at Austin. It is next to impossible for someone conducting a synthetic study to be intimately familiar with each of the sites from which information has been reported. In order to gain as much insight as possible, therefore, I sought and received invaluable information from people who actually participated in the original, site-specific studies. I owe, therefore, a debt of gratitude to the following people who provided firsthand knowledge about prehistoric canals at various locales discussed herein: Thomas H. Charlton, Richard A. Diehl, Herbert H. Eling, Gloria J. Fenner, Melvin L. Fowler, Emil W. Haury, T. Kathleen Henderson, J. Charles Kelley, Michael E. Murphy, E. Logan Wagner, and Barbara J. Williams.

A special note of thanks goes to B. L. Turner II and James A. Neely, who not only were sources of information about the details of specific canal sites, but also acted as sounding boards off which I bounced what now seem to have been an infinite number of (sometimes half-baked) ideas, and Karl W. Butzer and Gregory W. Knapp for their comments on various parts of the manuscript. I also thank William T. Sanders and Robert C. Hunt who provided insightful reviews of the entire text and argued, though in vain, for more regional and ecological focus. Finally, I thank Carol Vernon, and Beverly Beaty-Benadom who not only typed this manuscript but also tolerated my making repeated changes, John V. Cotter for drafting the maps, and, especially, my son David who compiled the bibliography.

1. Prehistoric Irrigation, Technology, and Mexico

Understanding Technology

Studies of ancient irrigation systems have comprised a major component of the prehistoric research conducted in Mexico during the past three and a half decades. For the most part, each of these investigations has focused on the remnants of a particular set of canals and related features. Much information on irrigation technology has been collected and reported as a result of these individual endeavors. Lacking, however, is a study that synthesizes this information and assesses it in a coherent framework.

Such a study had long been infeasible and unnecessary, but today is both possible and needed. Until recently, there were simply too few individual studies to provide an adequate data base, and, indeed, the quantity of the actual data did not justify an endeavor of this type. Recently, however, there has been an increase in the number of individual projects, and, accordingly, there exists a need to analyze the now abundant data base comprehensively.

In order to understand the nature of the individual systems being investigated, researchers have typically relied on comparisons with systems discussed in other studies. At one time there were only a few studies from which comparisons could be drawn. Assessments, therefore, tended to be all-inclusive, but they suffered because of the paucity of adequate comparative data. Today, studies are so numerous that detailed investigations of any one system usually include comparisons with only one or two others. These comparisons usually focus on either the temporal contexts, including systems of approximately the same age, or ecological contexts, including systems in similar environmental settings. Both approaches have their strengths, but they also have shortcomings.

Temporal Context

A number of studies involve comparisons of technology with only those systems that date to the same prehistoric period. Such studies provide much insight into the particular system being studied, but they are somewhat misleading, as differences between systems tend to be glossed over. The tendency is to emphasize similarities rather than differences. For example, in her assessment of early canals at one site in the Basin of Mexico, Deborah Nichols (1982a: 141) notes that "the levels of technology . . . required to construct and maintain this type of irrigation system are relatively simple . . . and would have been well within the range of activities undertaken by other Formative period communities." To support her claim she cites studies of Formative canal sites in the present-day states of Oaxaca, Tlaxcala, and Puebla (Nichols 1982a: 142).

A detailed investigation of all the systems cited (see Chapter 2) reveals that Nichols is absolutely correct. The system she studied was characterized by a level of technology well within the range of technologies exhibited by other sites dating to that general period. However, as is typical of studies that focus on the temporal context, Nichols overlooks two subtle, but nevertheless important points. The systems used for comparative purposes were not as contemporaneous, nor were the technologies as similar, as readers might be prone to believe. The system cited from Oaxaca is at least two hundred years, and the one from Puebla is no less than five hundred years younger than the one studied in the Basin of Mexico. Furthermore, the system reported from the basin was technologically much less complex than those cited from the other locales.

Ecological Context

Many studies do, of course, recognize the differences that exist between the technologies of various irrigation systems. Typically, these studies treat differences as functions of variable environmental conditions, and give no consideration to the dates or ages of the respective systems. Indeed, as in the case of Joseph W. Hopkins III's (1984) study of a relict canal irrigation system in the Cuicatlan Cañada of Oaxaca ca. A.D. 1000, there is a tendency to rely on ethnographic analogs using present-day practices in the same locales as parallels.

Today, perhaps more than ever before, prehistorians are concerned with the methods and techniques ancient people used in adapting to their local bio-physical environment (Butzer 1982). This cultural ecological perspective is predicated on the notion that people use whatever technology is needed in order to successfully inhabit a given area. For example, small populations that lived in narrow canyons incised by rapidly flowing permanent streams with steep gradients typically irrigated small plots by simply diverting water out of the channels upstream of their fields, allowing it to flow under the force of gravity through short canals (e.g., Redmond 1983, discussed in Chapter 3). Large populations that occupied broad, nearly flat floodplains of valleys with ephemeral streams, in comparison, usually needed, and hence used, intricate networks involving either water-raising devices such as diversion dams or long canals (e.g., DiPeso 1974, discussed in Chapter 4).

The exact ways in which the technology exhibited in such different systems was developed are, as a rule, not discussed. Instead, it is simply assumed, at least implicitly, that farmers devised the most appropriate way of irrigating a sufficient amount of land to satisfy the needs of the population. Such studies, for the most part, emphasize the adaptiveness of local farmers. Technologies are, therefore, commonly regarded as local, autochthonous innovations. Only recently has it been demonstrated how such adaptations could have taken place as the result of diffusion (Denevan 1983).

While ecologically based studies do offer different and additional insights lacking in studies that focus on only the temporal context, they also have their limitations. Specifically, by downplaying the importance of the ages of the respective systems and by overlooking the role of diffusion, such studies tend to treat the development of technology rather superficially.

Developmental Context

Studies that either compare ancient irrigation systems in their temporal contexts or emphasize the environmental factors responsible for the use of different systems fail to recognize the long-term and cumulative nature of technological change. Carl O. Sauer's (1952:9) observation that "Ideas must build upon ideas" seems to have been lost.

For the most part, technological change involves degree rather than kind. Although changes in types of technology have never been uncommon, changes in the nature of a particular kind of technology have been much more typical. For example, ancient farmers in the Valley of Oaxaca changed from pot irrigation—involving water extracted manually from shallow wells—to canal irrigation only once, but they made changes in canal technology numerous times (Kirkby

1973). Such changes in degree typically require that the users be both amenable to new ideas and innovative (Ford 1977 : 140–141). Irrigation technologies used in one environmental setting can rarely be diffused into a different one without some form of modification, which is either invented locally or diffused in from yet a third area. There is no way, for instance, that canal irrigation could have diffused from narrow canyons such as those cited previously and adopted on broad floodplains without the creation, or the introduction from elsewhere, of water-raising technology. Regardless of the specifics, the combining of the two technologies resulted in a more complex irrigation system that is quantitatively and qualitatively different from its predecessors. Because such change involves both an increase in size and an elaboration of existing phenomena, thereby resulting in a more advanced state, it can, by definition, be considered "development."

Recognizing technological differences in their developmental contexts has important implications for understanding the cultures that built and used prehistoric canal irrigation systems (Price 1971 : 9; Steward 1977 : 97; Hunt and Hunt 1974 : 131). For example, it has long been claimed that ancient cultures in the American Southwest were influenced greatly by people from Mesoamerica. Part of the evidence mustered in support of this proposition involves canal irrigation. According to the prevailing theory (Armillas 1949 : 91; Haury 1976 : 149), Mesoamericans must have introduced irrigation into the American Southwest because the oldest known canals in the latter area date to no earlier than 300 B.C. while those in the former area are much older. The argument is logical enough at first glance. It is weakened considerably, however, when technologies are compared in their developmental contexts. As will be elaborated in Chapter 3, canal irrigation systems in Mexico prior to 300 B.C. were smaller and technologically less complex than those in the American Southwest. If it is assumed that small, simple canal irrigation systems existed earlier than, and were necessary for the development of, large, complex ones, then the model that envisions Mesoamericans introducing irrigation and, hence, influencing Southwestern peoples to any great extent appears to be inappropriate.

Purpose of the Study

Technological developments have been taking place since time immemorial. This is perhaps no more evident than in the case of canal irrigation technology in prehistoric Mexico. Studies of individual relict systems indicate that numerous technologies of varying degrees of complexity were developed and used at different times, in different places, and in different environmental settings. These ranged from simple systems that used small canals to divert runoff from ephemeral mountain streams at an early date to elaborate networks that involved large canals to irrigate broad valley floors with water from perennial rivers, at the end of the prehistoric era.

This study reconstructs the sequence through which canal irrigation technology was developed. It does so by synthesizing data from individual studies and assessing them in their temporal as well as ecological contexts. As a result of this investigation, future investigations of individual systems will have a more accurate basis from which comparisons of individual canal irrigation systems can be drawn. Perhaps more important, however, this study adds to our general understanding of ancient Mexican peoples and their accomplishments.

Sources of Information

To say that an abundance of information exists on prehistoric irrigation in Mexico is an understatement. Indeed, it often seems as though every prehistorian working there has had something to say on the topic, at one time or another. Studies from which data on the technology of specific systems are extracted and analyzed in a developmental context fit into three categories. First, there are studies that do not present any direct data on irrigation but include inferences drawn from settlement data and ethnographic parallels. These are the most numerous of irrigation studies, but they also are the least reliable in terms of accuracy. Second, there are studies based on documentary accounts or historical records. These tend to be few, and limited to the period of Spanish contact. They are, however, reasonably accurate. Finally, there are archaeological investigations that actually provide detailed data on irrigation. These are the least numerous of the three types of sources but they provide the most information on ancient irrigation technology.

Analogs

Most prehistorians working in Mexico would like to find concrete and irrefutable evidence of irrigation. Relatively few, however, actually do. Although the absence of evidence is sometimes interpreted as evidence of absence, it often is not. In many cases there is reason to suspect that irrigation was practiced in a particular locale during prehistoric times but material evidence to that effect no longer exists. For example, some archaeologists believe that people living in the Valley of Oaxaca ca. 300 B.C. diverted floodwaters onto valley-bottom fields by means of small canals (e.g., Flannery and Marcus 1976:378). There is, however, no direct evidence of irrigated-related features dating to that period in that locale. The conclusion that irrigation was practiced there is based in large part on inferences made from the spatial distribution of sites inhabited at the time, and also on present-day traditional agricultural practices and assessments of the biophysical environment.

Making inferences about ancient irrigation on the basis of ethnographic parallels and settlement data is a common practice among prehistorians working in Mexico. It is, however, a risky practice, and interpretations can easily exceed the bounds of prudence. Accordingly, conclusions made on such grounds cannot always be taken at face value. Unless the number, sizes, and locational characteristics of the present-day settlements are nearly identical to those of the ancient sites dating to a particular time period, such inferences should be treated with caution. Furthermore, even if the nature of the irrigation can be inferred with reasonable accuracy, the technology involved can only be surmised. In this study, inferences made about ancient irrigation in Mexico on the basis of analogs are used sparingly, only when other forms of data are not available, and then only after careful scrutiny.

Historical Accounts

When Spanish soldiers, missionaries, and government officials arrived in the sixteenth century, they noted that irrigation was being practiced at literally hundreds of locales in Mexico. Their observations are replete with statements such as "irrigation is practiced there" and "canals carry water to their fields." These references provide a great deal of information about the extent to which irrigation was being practiced, but they rarely contain information about the technology itself. Furthermore, they indicate only what existed at one very late and brief period.

It is only with the very large and complex canal systems in use at the time of contact that documentary relations provide much insight into prehistoric canal technology. Some features, such as large aqueducts, not only were conspicuous to the Spaniards but were sufficiently spectacular to merit detailed descriptions. A few studies have gleaned data on technology from these historical documents, and these are relied upon occasionally herein.

Archaeological Reports

Without doubt, the best sources of data on prehistoric canal irrigation technology in Mexico are the numerous books, articles, and unpublished reports that discuss the findings of archaeological projects in which material remains of irrigation systems have been confirmed during the course of fieldwork. Frequently, but certainly not always, identified features have been mapped, excavated archaeologically, dated, described, and illustrated.

Almost paradoxically, there is at once both a paucity and a great deal of archaeological data on prehistoric canal irrigation technology in Mexico. Confirmed evidence has been found in relatively few locales (Fig. 1.1). Although not abundant, the data from these places are of high quality. Most archaeological researchers have been so intrigued by their discovery of a canal network that they have expended a great deal of energy in both recovering and reporting the data. In addition to the usually quite thorough initial investigations, there also have been several follow-up investigations or restudies of canals, often by scholars other than those who originally conducted research at the various sites. The end result of the discovery of an irrigation system, therefore, is typically the publication of more than one report or article, by more than one person. Although there are few places with confirmed evidence of canals, there have been a large number of investigations, resulting in a yet larger number of publications.

Unfortunately, however, in some cases the interpretations of how and when various systems functioned can be misleading. For example, linear features purported to be remnants of canals comprising an irrigation system have been reported approximately 7 kilometers up the Tecolutla River from the Gulf of Mexico on the coastal plain of northern Veracruz state. Assessed principally through the use of aerial photographs and surface surveys, these so-called "canals" are thought to have been associated with the ancient Totonac site of El Tajín, and therefore dated between A.D. 300 and 1000 (Wilkerson 1980:217–219). Although they are striking and were clearly agriculturally related (Wilkerson 1983:59–78), there is no way that these features can be considered irrigation canals in the true sense of the term. They were not constructed in order to bring water to previously dry lands. These features are located in a rela-

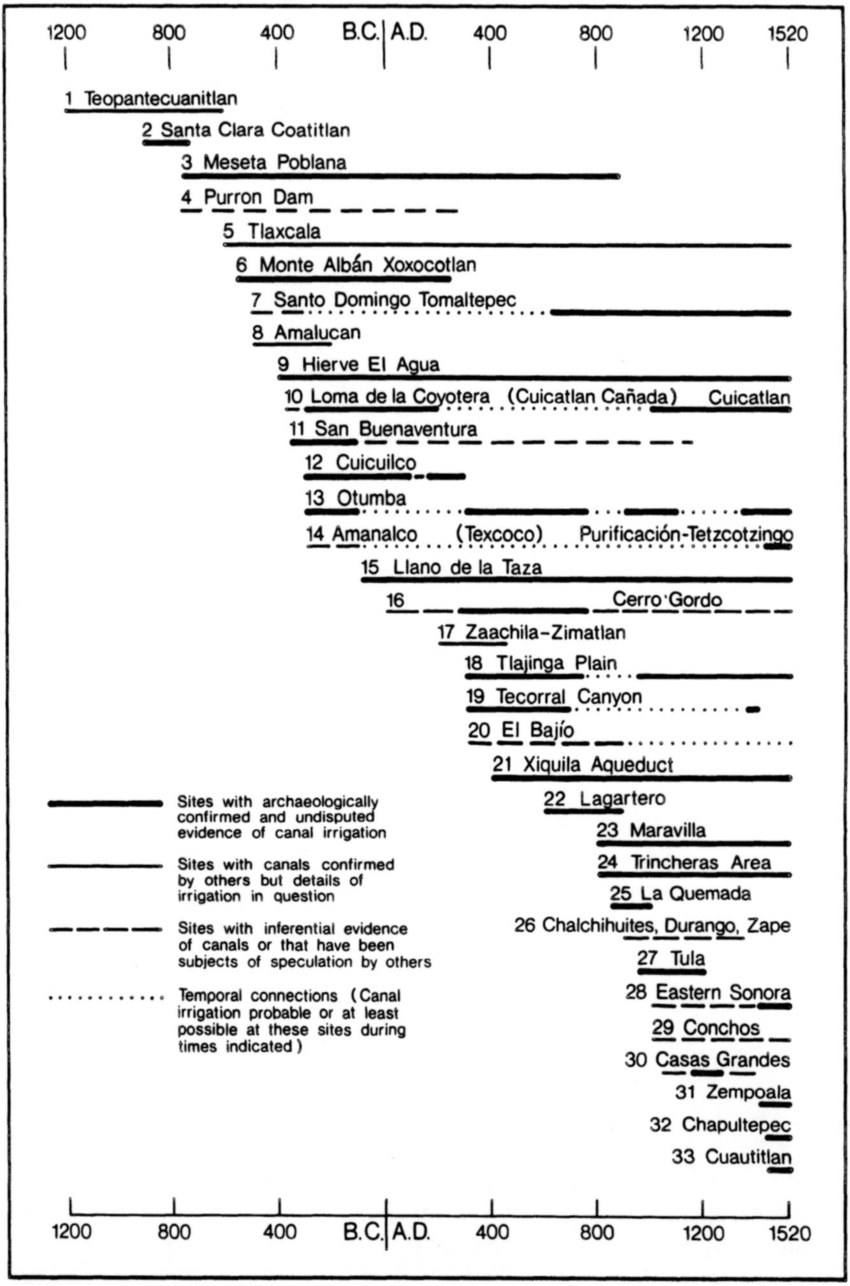

Figure 1.1. Data on prehistoric irrigation canals in Mexico.

tively relief-free deltaic area that is seasonally inundated. In all probability, they were previously natural distributaries that were channelized for a raised field, wetland-reclamation project (Siemens n.d.). No other prehistoric irrigation canals have been found in coastal northern Veracruz, but abundant evidence of raised fields exists in the regions (e.g., Siemens 1983). Also, although canals have been noted at numerous Maya sites, "There is little evidence of . . . irrigation as commonly conceived" having been practiced anywhere in the lowlands of eastern and southern Mexico (Siemens 1987 : 668).

In other cases, problems in the manner in which the evidence is reported cast reservations on the interpretations. For example, extensive irrigation works that were claimed to have been used for over two millennia and involved numerous technological improvements have been reported from the highlands of present-day Tlaxcala state (Abascal and García Cook 1975; García Cook 1981; 1986). Although such accounts are tantalizing, few data are provided that are sufficiently detailed to be usable. To date, reports from the area contain no indication that the purported systems were studied systematically, much less tested archaeologically. Conspicuous by their absence are maps, photographs, illustrations of various features, and profiles of excavated canal cross-sections. Equally deficient are details concerning dating and chronology. Various irrigation systems have apparently been classified intuitively on the basis of technological complexity and then fitted into the chronological sequence established for the region. Simpler systems were arbitrarily assigned to the earlier phases and more complex ones to the later phases (William T. Sanders, personal communication, 1987).

The above example illustrates that archaeological data must be carefully scrutinized before they can be accepted as reliable. For the most part, however, assessments made by the people responsible for the discovery, analysis, and interpretation of the various systems are very good. Even in those cases where interpretations are less than perfect, the data themselves are usually more than adequate as long as they are presented in detail.

Lending credibility to most interpretations, is the degree to which the reported ancient irrigation systems have been preserved. Most such canals and other irrigation-related features have been only moderately affected by historic and recent activities. With few exceptions, the majority of known prehistoric canals are in areas that have not been densely populated since the respective irrigation systems were abandoned. In those cases where relict canals are located in well-settled places, the present-day people are either no longer engaged in farming, or, if they are, a type of irrigation other than that involving canals is being used. In the former case, such as at Cerro Tetzcotzingo, people have no need to use, and hence disturb, areas that were formerly involved in irrigation. In the latter case, such as on the Llano de la Taza, the lands irrigated prehistorically may still be cultivated, but the water used for irrigation is typically pumped out of deep wells. Under such conditions, many formerly used canals remain unused and undisturbed.

That many ancient canals remain largely intact should not be construed to mean that they all do. In speaking of hydraulic features in the Tehuacan Valley, for example, Richard B. Woodbury and James A. Neely (1972:100) note that those reported are not "unique; instead they are all that have both survived and been recognized." Neither should one think that extant features are not in danger of being destroyed. Many areas that were intensively irrigated during prehistoric times are now densely settled, and the population is continuing to grow and occupy more land. The premier example, of course, is the Basin of Mexico, the location of modern Mexico City. There, urban sprawl not only has destroyed many canals that were never recorded, but has removed all traces of canals at one known site—Cuicuilco—which were the subject of investigation in the late 1950s, and is threatening others, most notably the canals at Santa Clara Coatitlan (Fig. 1.2).

It should not be thought that no prehistoric canal sites other than those discussed here exist in Mexico. There are probably several that either are known to local inhabitants but have not been brought to the attention of archaeologists or are known to archaeologists but lack proper study and reporting. Furthermore, there may well be thousands that have yet to be discovered, or as Woodbury and Neely (1972:100) put it, "recognized" by anyone. Indeed, even near Mexico City, previously unknown prehistoric canals, for example those of Tlajinga near Teotihuacan, have been found only within the past few years.

Known prehistoric irrigation canals in Mexico not only are reasonably well preserved and well described, but also have three other qualities that make them most acceptable for a study of technological change. First, the cumulative data for these systems span a very long period of time, over 2,700 years, therefore providing a good temporal context. Second, extant canals are located in environmentally diverse locations scattered over a wide area, thereby providing good spatial and ecological context. Third, with few exceptions, each canal or canal system was used for a relatively brief period thereby facilitating understanding of how development took place.

Although archaeologists normally prefer to study sites that were

Figure 1.2. Valley of Santa Clara Coatitlan, where traces of prehistoric canals are rapidly being obliterated by the expansion of Colonia Xaltostoc, a suburb of Mexico City.

occupied for extensive periods, there is a clear advantage for a study of technology to have data that come from sites used for only brief periods. In such cases, remnants of the specific items under investigation, specifically irrigation canals and related features, tend to be preserved in much the same form as when they were operated by their builders. Features used over lengthy periods frequently are subjected to numerous modifications that obliterate many traces of earlier conditions. Because they were not used very long, most of the prehistoric canal systems in Mexico were not modified to any great degree by later irrigators.

Terminology of Canal Irrigation Technology

Information on prehistoric canal irrigation systems in Mexico is extensive, and the confirmed data are detailed. Indeed, there is hardly a case in which morphological details of irrigation features have not been described meticulously once they were found. Prehistorians' deep and inherent concern for recovering all types of ancient data is reflected in their careful descriptions of canals and related features.

Unfortunately, many irrigation features were recorded by individuals who know very little about water control, even though they appreciate its importance. The lack of experience some investigators have with irrigation has resulted in a variety of terms being used to describe similar features. For example, the terms "canal," "ditch," "channel," and even "aqueduct" have all been used in discussing functionally identical earthworks. Such inconsistency can cause confusion for a comparative analysis.

There is also some confusion as to what is actually involved in canal irrigation. The literature on ancient agriculture in Mexico is saturated with references to all sorts of features considered by their discoverers and reporters to be parts of "irrigation systems." As in the case of El Tajín, some canal-like features may have been associated with agriculture but not with canal irrigation per se. Given both the vagueness of and the uncertainty over the nature of the various systems reported to date, it is essential that the mechanics of canal irrigation be clearly outlined. Furthermore, because of the variety of terms used in reference to specific features, it is imperative that a basic set of terms be defined before any further discussion ensues (Hunt and Hunt 1974:131).

Irrigation is the artificial application and distribution of water to otherwise dry lands in order to facilitate cultivation (Monkhouse 1965:173; Moore 1974:117). As its name so accurately indicates, *canal irrigation* is a specific form of irrigation that involves the transport of water from a source by means of gravity flow through artificially constructed open conduits or canals. The terms "perennial irrigation" and "wadi irrigation" have been used to distinguish between those systems that involve constantly flowing and ephemeral or seasonally flowing sources, respectively (N. Smith 1975:5, 6). Although applicable in the Old World, these terms have problems that make them inappropriate for use in regard to Mexican irrigation. First, the term "perennial" also applies to situations in which "Irrigation is practiced all year-round. Different crops can be cultivated at various times of the calendar" (N. Smith 1975:229). Second, the term "wadi" is applied to very long and broad-bottomed, typically dry stream channels in the Middle East. Although the term "arroyo" is frequently used synonymously in the New World, it is not quite the same—wadis are notably larger and possess different hydrologic characteristics; specifically, they often do not empty into other streams but rather terminate in the desert (Moore 1974:16–17, 236). The terms *permanent irrigation* and *floodwater irrigation* are more applicable to the situation, and hence more commonly used, in Mexico (Turner 1983). They are, accordingly, used here. Regardless of the water source, however, canal irrigation systems have

Table 1.1. *Features associated with prehistoric canal irrigation systems in Mexico (by section)*

Headwater Features	Canal Features	Field Features
Channelized streams	Aqueducts	Field canals
Dikes	Main canals	Drainage ditches
Weirs	Distribution canals	Water spreaders
Diversion dams	Lateral canals	Bunds
Storage dams	Head gates	Bordered checks
Spillways	Sluice gates	Furrows
Floodgates		Terraces

three distinct sections (Rydzewski 1987:229), each characterized by particular sets of features (Table 1.1).

Headwater Features

Furthest upstream are found devices used to control the flow of water through or out of its *channel* or natural stream bed (Monkhouse 1965:59; Moore 1974:34). In some cases, channels have been enlarged or rerouted. These streams are considered to have been *channelized* (e.g., Jansen et al. 1979). In other cases, channels have *dikes*, wall-like structures built along their banks in order to reduce stream meandering and flooding of adjacent lands (e.g., Winkley et al. 1984:101–103).

A number of different terms, including "dam," "barrage," and "weir," have been applied to features built across channels at the heads of irrigation systems. Although "dam" is undoubtedly the most commonly used, and its general meaning is widely understood, the term is not without difficulties. The greatest problem is that dams function in a number of different ways. Depending on the circumstance, dams either impound or store water, raise the surface level so that water can flow into canals, or assist in the diversion of water. In some cases they perform a combination of impounding, raising, and diverting. Perhaps not surprisingly, numerous terms have been used in discussing various types of dams according to their function. Much more clarification is needed.

For the sake of simplicity, standard hydraulic engineering terms are the most ideal to be used in reference to these irrigation-related features that are built across streams. *Storage dams* are devices intended principally for impounding water. They create reservoirs, storing surplus water that can be used during rainfall-deficient periods (Arthur 1965:63). Excess water does not normally flow uncontrolled over their tops, but instead is drained through either a *spillway*, a passageway near the top of the dam (N. Smith 1971:266), or a *floodgate*, a passageway near the bottom. Devices that are built only to raise the level of water to a height where it can empty into a canal are known as *diversion dams*. Although they impound to a certain extent, these features are not intended to store water. There are no spillways or floodgates on diversion dams (Arthur 1965:63). Instead, water flows over from their tops, in some cases constantly. In some parts of the world, diversion dams are also known as barrages, although the term is most commonly used for large-scale impoundments (N. Smith 1975:227).

The onset of flow in ephemeral streams is sometimes so rapid that the terms "flash flood" and "wall of water" are commonly used to describe such events. Although applicable in some extreme circumstances, the onset of flow can also be slow, with small amounts of water confined to the bottom of channels. In such cases, the initial head of water tends to follow the natural course of the stream even if the mouth of a canal intersects it at the same elevation. Accordingly, *weirs* are often constructed across channels in order to divert the flow of water into canals. Because ephemeral streams eventually flow quite rapidly and with great force, weirs normally function only briefly. They are temporary structures, often built only of brush, that have to be rebuilt regularly.

Canals and Related Features

Through their mid-sections, canal irrigation systems are characterized by a single large canal. Because they begin at the source and carry virtually all of the water used in a given system, these features are appropriately referred to as *main canals*, supply canals, or principal canals. In Mexico, as throughout the Spanish-speaking world, a canal of this type is known as an *acequia madre* or mother canal, as it bears vital water (Meyer 1984:36).

The middle sections of canal systems can contain four other types of features—aqueducts, head gates, sluice gates, and branch canals. Some writers use the term "aqueduct" as a synonym for "main canal." Although the two terms are technically interchangeable (N. Smith 1975:227), *aqueduct* commonly has a more specific definition—it is an elevated structure used to maintain an appropriate gradient for the flow of water over broad, low-lying areas or narrow but deep ravines (Prager 1978). In some situations aqueducts are simple earthen features. In others, however, they can be bridge-like devices.

Controlling the amount of water entering and exiting main canals is no easy task, especially in permanent irrigation systems where constantly flowing water must be handled. Regulating water flowing through a given system at any one time requires the use of devices that can be opened and closed to varying degrees at each end of the canals. Such features can take on a number of different forms. Generically, however, those at the far upstream end of the main canal, near the point where water is directed out of the stream channel, are referred to as *head gates* (Campbell 1986:17). Those at the junction of any two canals are called *sluice gates* (I. H. Adams 1976:117).

Toward the downstream end, the main canal typically feeds water into a series of smaller canals. These have been referred to variously as "branch canals," "secondary canals," "lateral canals," and even "feeder canals." Each of these terms is acceptable. They are not, however, interchangeable. Those that carry water to and terminate at the fields are the *lateral canals*, which typically run nearly perpendicular to the main canal. *Distribution canals* are those that carry water from the main canal to the lateral canals. In the most general sense, all of these small canals at the lower end of the canal system can be considered *branch canals*.

Not every system, of course, has each type of branch canal. In general, larger systems have more numerous and more varied types of canals than smaller systems. Typically, the smaller the system, the less likely lower-level distribution canals will be found. For example, in the smallest irrigation systems known, main canals are all that exist; these run straight from channels to individual fields. Larger systems often have distribution canals that terminate in fields, and still larger ones have laterals that water fields directly. In effect, the more numerous and varied the canal types, the more technologically complex the system.

Field Features

Once water arrives at the fields, a number of devices are used to distribute it across the planting surface. *Field canals*, as their name implies, carry water onto and across field surfaces.

Controlling water on the fields is typically not much easier, and in some cases can be more difficult, than getting it through the canal network. Essentially, there are only two problems that are encountered, distributing the water evenly over the field surface and ensuring that it does not run off too rapidly. *Water spreaders* (French and Hussain 1964) are used in some cases to distribute water evenly across the gently sloping surface. Typically made of either brush woven among stakes driven into the ground, earth, or rocks, water spreaders are low and rarely more than a few meters long. They are oriented across the slope or perpendicular to the general direction of the flow. Although they are often found scattered across a field or in groups, water spreaders are almost always located in gullies where surface flow is concentrated.

In addition to providing every part of their fields with water, irrigators are also faced with the problem of ensuring that the water stays on the fields long enough to wet the soil sufficiently. *Wild flooding*, the practice whereby irrigation water is allowed to flow across and off the fields (Israelsen and Hansen 1962:297–299; Rawitz 1973:323–324), is sufficient in some cases, especially where irrigation water is abundant. In other cases, however, irrigators cannot afford to water their fields in such an uncontrolled manner. The most traditional method of retaining water on fields that are nearly flat is by constructing mounds of earth around the perimeter of the cultivated area. Such features have been referred to variously as "berms," "earthen embankments," "linear mounds," and even, incorrectly, "dikes." The literature on agricultural engineering refers to them more specifically as *bunds* (Krantz and Kampen 1979:282–285).

Large fields sometimes have numerous bunds oriented across the slope. In some cases bunds even crisscross planted areas, creating a series of *bordered checks* (Israelsen and Hansen 1962:299–302; Rawitz 1973:323–325). Obviously, the more bunds on a field, the better the control of water.

Another method of controlling water on fields involves the use of *furrows*, series of ditches usually less than a meter in width and depth, separated by ridges that are equally wide and high (Israelsen and Hansen 1962:303–305; Zimmerman 1966:107–130; Rawitz 1973:325). Furrows typically, but certainly not always, run parallel to the contours of the land surface and perpendicular to the slope. In a sense, they are tantamount to a large number of field canals. Similarly, the ridges between them are the equivalent of a dense network of bunds.

Getting the excess water off fields is a problem in some cases, and this is typically accomplished by means of ditches. As has been stated previously, "ditch" is often used synonymously with "canal." There are, however, differences between the two types of features. The term "canal" connotes an element of formality and care in excavation that is not always associated with the term "ditch." Ditches are usually crudely constructed, and, indeed, they are often the by-products of other building efforts (e.g., "borrow" or "bar" ditches). Canals, in contrast, are constructed with much attention to details. Also, whereas canals carry water to a place, *ditches* are used to collect water and transport it away from a locale, as in the case of draining excess water off a field (Hillel 1987:53). In its verb form, "ditch" does, in fact, mean to discard. In the context of dry-land irrigation systems, therefore, ditches are usually small, often rather crudely constructed features used commonly for draining fields. Exceptions in the size and quality of construction do, of course, exist. The function, however, is always drainage (I. H. Adams 1976:97, 126).

Canal irrigation, especially in Mexico, is typically found on low-lying, nearly relief-free plains. There are, however, cases in which slopelands are irrigated. In such places, runoff is a major problem that cannot be controlled by either water spreaders or bunds and ditches. Accordingly, *terraces*, artificially leveled surfaces, usually faced and supported by rock walls, are often built in order to facilitate cultivation (J. E. Spencer and Hale 1961:3; Donkin 1979).

In sum, canal irrigation systems are composed of a number of different types of features, each varying considerably in size and characterized by varying degrees of technological complexity. These features can be, and have been, combined in a seemingly infinite number of ways. Depending on the combination, canal irrigation can range from the simple diversion of ephemeral stream flow to a small single field, to the transport of water several kilometers, over numerous obstacles, through intricate networks of branch canals and onto elaborately prepared fields. Nearly every type or combination imaginable was used at one time or another in Mexico during prehistoric times. The existence of such variety in different locales at different times attests to a complex sequence through which the technology developed.

Premises

Two basic premises about the behavior of ancient farmers underlie this study. The first is that irrigators were pragmatic people concerned with solving immediate problems (see Wilkinson 1973; Ford 1977:140, 141), in this case growing a sufficient amount of food on lands with water deficits. These farmers were both inventive and open to new ideas that would ensure success. While they were un-doubtedly cautious and conservative, not willing to abandon successful practices, they were receptive to introduced techniques that seemed promising, and they were constantly experimenting (see Barlett 1980).

Solutions to their immediate problems were, of course, varied. Nevertheless, prehistoric irrigators in Mexico shared a second behavioral attribute—they were intent on meeting their implicit production goals in the easiest way possible (see Zipf 1949). Holding other factors constant, these farmers typically chose the least demanding solution to their agricultural problems. It follows, then, that the smallest, simplest, and least demanding technologies were used before large, complex ones (e.g., Boserup 1965; 1981). Developments in technology came about, therefore, only as demands increased and new problems arose or were encountered. As the saying goes, necessity is the mother of invention.

Organization of the Text

The body of this text is devoted to the detailed documentation of the evidence of canal irrigation in ancient Mexico, and to the analysis and interpretation of those data. Specific systems are assessed in terms of their chronology: the oldest system is examined first and the youngest last. Throughout the discussion, focus is placed on understanding the technological elaboration of each system over its predecessor. The spread of canal technology into and through different ecological settings is also traced. In effect, the sequence or trajectory through which canal irrigation technology developed is reconstructed.

Chapter 2 focuses on the earliest centuries for which data on ancient canal irrigation can be found in Mexico. It begins with a discussion of a system that some claim to be the oldest known, ca. 1200 B.C., and reviews data on other early systems that involved the collection and distribution of ephemeral runoff. The earliest spring-fed canal system is also discussed.

Chapter 3 begins with an assessment of the first attempt at diverting water out of perennial streams, ca. 300 B.C. How technology developed in order to irrigate broad valley bottoms is considered next. This chapter concludes with an examination of some isolated, and small, but nevertheless important, technological changes.

Chapter 4 deals with systems that were brought into use between A.D. 800 and 1200. The first section focuses on developments in far northwestern Mexico, on the border of the Greater American Southwest culture area. The second part assesses what might have been going on in terms of irrigation in that expansive and largely unknown area between the Southwest and Mesoamerican culture core.

The chapter concludes with discussion of technological accomplishments in Mesoamerica.

Chapter 5 deals comprehensively with achievements in canal irrigation technology during the era immediately preceding the arrival of the Spaniards. Discussed are developments throughout the region as well as some truly monumental accomplishments that occurred in the Basin of Mexico.

Chapter 6 discusses the origins as well as some of the implications of the developmental sequence of canal irrigation technology delineated in the previous four chapters. Two possible scenarios for its origins are outlined. The nature of the developmental sequence as a whole is reviewed in terms of its importance to understanding changes in the organization of "hydraulic societies."

Finally Chapter 7 focuses on some of the more important individual elements of the technology developed prehistorically. The introduction of new technologies by the Spaniards and later Mexicans is also discussed. Of particular importance here is how prehistoric canals were renovated in order to continue to contribute to the county's agricultural economy.

It must be emphasized that this study is a documentation and assessment of the development of prehistoric canal irrigation technology in Mexico. It is not intended to explain why, or how, the ancient Mesoamerican high cultures arose. The study certainly pertains to this line of inquiry (e.g., Steward et al. 1955; Price 1971) and has specific implications of broader cultural interest. The overriding theme, however, is to better understand how the canal technology itself was developed.

2. An Era of Experimentation, 1200–350 B.C.

Development of Floodwater Systems

Understanding the genesis of canal irrigation technology has been difficult, largely because of the rapidly changing data set. About the time one canal system was accepted as the earliest, a yet older one was found. The last few discoveries have, however, one thing in common: the canals all involved the diversion of water from ephemeral streams.

Teopantecuanitlan

At the present time the oldest features in Mexico that are claimed by some to be a canal irrigation system are associated with the recently discovered Olmec site at Teopantecuanitlan near the confluence of the Ríos Mezcala and Amacuzan in the far northern part of Guerrero state. Because it was found only recently, little is actually known about this site. Information is in the form of one brief professional article (Martínez Donjuán 1986), press releases about the discovery (e.g., Crossley 1986; Vargo 1986), and relations from people who have visited the site during excavations. Although minimal, data do provide some insight.

The site was established perhaps as early as 1400 B.C. on one side of a natural channel that collects ephemeral runoff from a high hill behind it. A feature thought to be a storage dam was built across the drainage sometime later, most likely between 1200 and 1000 B.C. (Crossley 1986). It is reported that this feature was built of both rough, uncut stone, and faced masonry blocks (Wagner 1986). It is estimated to be 30 meters long, 3 meters high at the center, and resulted in a reservoir covering an area measuring approximately 20 by 30 meters. If descriptions are correct, it appears that this feature can be considered a "gravity dam," one that is straight, or perpendicular to the stream flow, and resists the applied water load by means of its weight (N. Smith 1971 : 265). Although no spillway has been reported, it seems likely that some sort of gap or low point would have been built into the top of the dam, probably near the end opposite where the canal begins.

The feature identified as a "canal," and supposedly leading away from the dam, is over 300 meters long and terminates in what has been described as fields covering a few hectares on the floodplain below the site. It varies in width from 70 to 90 centimeters and in depth from 90 centimeters to 1.4 meters (Martínez Donjuán 1986: 64). Both the sides and the bottom of the canal are lined with very large, thick, unfinished rock slabs. Only about 100 meters of its length has been exposed through archaeological excavations. It is thought, however, that this canal might be joined by at least two morphologically similar features that drained the central plaza of the site itself, much as in the case of the well-studied and famous Olmec site of San Lorenzo in the present-day state of Veracruz (Coe 1968:41–71).

In San Lorenzo, twenty small lagoons, thought by some scholars (e.g., R. E. W. Adams 1977:84) to be water storage ponds and by others (e.g., Coe and Diehl 1980:30) to be ceremonial baths, were drained by a complex of small conduits that led into one larger conduit. These drainages were constructed by linking end to end a series of short U-shaped basalt troughs. The main drain was 170 meters long, 15 centimeters deep, and 26 centimeters wide on the inside. The entire system was covered with flat basalt lids (Coe and Diehl 1980:118,120). The walls and the bottom of the troughs were quite thick, indicating that carving skills were minimal. Also, differences in workmanship indicate that a number of stone carvers were employed and that "control was lax during construction" (Coe and Diehl 1980:122). The system was built between 1000 and 900 B.C. (Coe 1981:122).

The interpretation of the rock feature as a storage dam and the stone-lined canal as being for irrigation purposes at Teopantecuanitlan seems plausible enough. It should not, however, be accepted without some skepticism until further research is done at the site. There are four reasons for this. First, the features most similar in morphology to the "canal" at Teopantecuanitlan—the drains at San Lorenzo—were not built until at least two hundred years later. Second, the drains at San Lorenzo were not associated with agriculture, much less used for irrigation (Coe and Diehl 1980:389). Third, the size and nature of the "dam" combined with the relative elaborateness of the "canal" are not in congruence with other evidence of early irrigation. At the risk of getting too far ahead too quickly, there is abundant evidence of technologically less complex systems later in prehistoric times and no evidence of systems comparable to this one having been built until several centuries later. Finally, another early agricultural system set in a locale much like that at Teopante-cuanitlan had its initial "irrigation" interpretation overturned by further studies.

Sometime between 1000 and 750 B.C., a group of "hydraulic con-structions" were carried out at the site of Chalcatzingo in what is now the state of Morelos. When first reported, these works were thought to have comprised "small-scale irrigation systems," in part because of the existence of so-called "terraces" and "diversion dams," one of the latter of which was 35 meters long and 7 meters high (Grove et al. 1976:1208, 1205). When the final analysis of the site was completed nearly a decade after the initial interpretation was made, however, it was clear that the terraces were "unirrigated" (Grove 1984:24). Furthermore, the alleged "dams" neither diverted water onto the terraces nor impounded it. These features, as it turns out, served to direct water around sharp curves in the natural chan-nels, minimize lateral erosion (meandering), and keep unwanted floodwaters off the terraces (Grove 1984:45). In effect, they were dikes (see also Grove and Guillén 1987:32–33, 41–42; Prindiville and Grove 1987:79).

Santa Clara Coatitlan

The earliest undisputed evidence of canal irrigation in Mexico comes from Santa Clara Coatitlan, located in the northern part of present-day Mexico City. Found by members of William T. Sanders' Basin of Mexico survey team during the inspection of two modern clay ex-traction pits in 1974, these canals were originally thought to be con-temporaneous with occupation at Teotihuacan (Sanders and Santley 1977). Reinvestigations, however, found that they actually dated to the early part of the first millennium B.C. (Nichols 1982a).

This system involved a channelized gully that collected tempo-rally erratic runoff from the eastern slopes of Cerro Guadalupe be-fore it emptied into Lake Texcoco. This channel was slightly more than 2 kilometers long (Fig. 2.1), with a relatively steep gradient of approximately 1 percent. According to the more recent and accurate account (Nichols 1982a:137), it was approximately 1 meter deep and a little more than 1 meter wide across the top. It had a trape-zoidal cross section with rounded corners, with the bottom slightly narrower than the top. No evidence of a dam has been found.

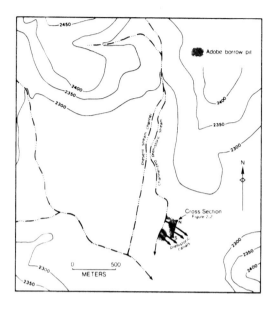

Figure 2.1. Map of canals at Santa Clara Coatitlan. After Sanders and Santley 1977.

Toward its downstream end, a series of at least twenty-five small U-shaped, unlined canals, each averaging approximately 60 centimeters wide and 50 centimeters deep, branched off the channel at nearly right angles (Fig. 2.1). These canals took water directly to individual fields (Nichols 1982a: 139). Although the maximum area cultivated was probably between 10 and 20 hectares, it is not likely that more than a few hectares were irrigated at any one time. Details are lacking because of widespread destruction resulting from present-day urban activities, but it appears that fields were flooded simultaneously.

There is no evidence that any type of diversion devices were used in order to direct water out of the channel and into the canals. The first investigation of the site concluded that some type of "dam" was "probably" used to raise the level of the water in the relatively deep channel so that it flowed into the relatively shallow canals (Sanders and Santley 1977 : 586). In all likelihood, however, this interpretation is incorrect. Water lifting and diverting devices were probably neither needed nor wanted.

From an analysis of an enlarged topographic map of the area, it appears that prior to construction of the canal system, the area cultivated prehistorically was inundated with sheetflow off the Guadalupe Range every time it rained. Rather than being excavated as a canal that brought water to the fields, the gully might well have been channelized in order to contain and divert damaging flood-waters around and away from the cultivated area. Because the channel is deeper than the canals, it probably carried more water more often. The canals probably carried water only after exceptionally high runoff events. Evidence of earth and rock obstructions has been found in at least one of the canals (Sanders and Santley 1977 : 586). Presumably, such obstructions were water spreaders that helped control the flow of water over the fields.

The collection and containment of runoff from surrounding slopes in a channelized ephemeral stream bed might appear to have been an easy task. Although great amounts of labor, social organization, and skill were not needed to build this system, its users should not be viewed as having done little. Evidence suggests that the irrigators at Santa Clara Coatitlan were constantly struggling with their canals. The rather steep gradient of the channel resulted in both rapidly moving water and the deposition of coarse sediment, both sand and gravel. On at least one occasion, the channel had to be re-excavated (Nichols 1982a : 136, 139), and the canals were filled, abandoned, and redug numerous times (Sanders and Santley 1977 : 585) (Fig. 2.2). The canal system was initially constructed around 900 B.C. The problems with sedimentation and maintenance were apparently so great that it was used for less than two centuries. The system was abandoned around 725 B.C. (Nichols 1982a : 141).

The channel and the canals at Santa Clara Coatitlan stand as the earliest evidence of people attempting to control *active*, flowing water in Mexico. Unlike the purported canal at Teopantecuanitlan

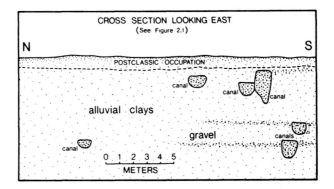

Figure 2.2. Cross section of distribution canals at Santa Clara Coatitlan. After Sanders and Santley 1977.

that carried *passive,* stored water that could be released from the reservoir at the farmers' convenience (if it functioned as some suspect), the canals at Santa Clara Coatitlan carried water only when an excessive amount was flowing off adjacent slopes. Such flow was almost completely unregulated. That these farmers were something less than successful in their attempts to use the runoff for agricultural purposes is evident by the relatively brief period of time in which the system was used. These irrigators eventually abandoned the system. Their efforts were gallant, however—and, perhaps more important, they had long-term implications in regard to the development of canal irrigation technology.

Meseta Poblana

The control of flowing water appears to have been mastered by irrigators on the Meseta Poblana, 10 kilometers east of the present-day city of Puebla. There, two types of canal systems have been identified through the use of aerial photographs and surface surveys. According to Prudence S. Precourt (1983 : 265, 267), who made the discovery, one type of system involved the diversion of water out of one ephemeral stream channel into canals, across intervening fields, and then into ditches that emptied into another channel parallel to the first one (Fig. 2.3). The second type of system is thought to have involved canals that carried water from a stream to a "green zone," an area in which local residents attempted to maintain high moisture levels (Precourt 1983 : 267).

Sufficient evidence does not exist to either confirm or refute Precourt's assessment of the second type of system. Suffice it to say that her interpretation seems sufficiently plausible. That the first type actually involved canals and ditches that connect natural channels, however, is untenable. Precourt (1983 : 269) interpreted one canal (feature 1 in Fig. 2.3) as flowing the wrong direction, probably because she did not depict the system on a topographic map. An assessment of her map after the contours were added revealed that, with the exception of two distribution canals (features 2 and 11 in Fig. 2.3) that take water out of another canal, the confirmed main canals either have their upstream ends at natural channels (features 1, 5, 10, and 12 in Fig. 2.3) or were identified only by isolated segments (features 3, 6, 7, 8, and 9 in Fig. 2.3). In not one case does Precourt have confirmed evidence of a ditch emptying into a second natural channel! In all probability, fields were located in those places where drainage ditches are claimed to have been. It can be concluded, therefore, that there existed only one type of canal system

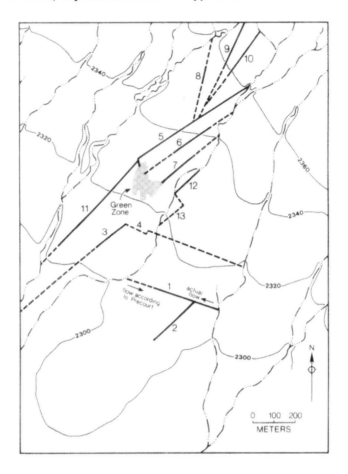

Figure 2.3. Map of canals at Meseta Poblana. After Precourt 1983, contours plotted from 1 : 50,000 topographic map, S.P.P. Puebla, E14B43.

on the Meseta Poblana. It involved main canals that transported water from channels into distribution canals that carried water to fields that could not have been more than 5 hectares each in area. The total area irrigated by all the canals was probably no more than 35 hectares.

That no evidence of dams was found at the head of any of the canals suggests that some type of temporary stake and brush weirs were probably used to help direct water into the canals. Such devices are about as basic as can be (Doolittle 1984a: 131). Weirs of this type are easily destroyed by floodwaters and have to be rebuilt regularly, at least annually and usually just before the onset of the rainy season. Because they are temporary, they never leave material remains necessary for archaeological confirmation. Although there is no confirmed evidence of weirs, they might well have been developed here. Certainly there is no earlier evidence for the use of such features elsewhere, and none of the earlier canal systems needed them.

Precourt's interpretations of both the regional hydrology and the irrigation system are somewhat erroneous. Her identification of the features as canals and her evaluation of the associated sites by which she dates them, however, are acceptable. Precourt appears to be correct in concluding that the canals were not all built at the same time. Evidence indicates that the first few (features 1 and 2 in Fig. 2.3) were built between the middle and the end of Nogal times, or sometime after 750 B.C. but no later than 300 B.C. They irrigated a maximum of 27 hectares (Precourt 1983:269). Although the remaining canals were brought into use over the next 1,500 years (Precourt 1983:332), none had changed in terms of technological complexity. Throughout the entire known sequence of prehistoric irrigation on the Meseta Poblana, canal technology remained most elementary.

Although simple, this system was a technological improvement over the canals at Santa Clara Coatitlan. That it functioned for a millennium and a half speaks for itself. There is more, however. The canals at Meseta Poblana comprise the earliest known system involving distribution canals that took water out of a main canal.

Purron Dam

At approximately the same time that canals were being built in order to divert floodwaters onto fields at Meseta Poblana, a storage dam was being built to retain the flow of an ephemeral stream further to the southeast in the Tehuacan Valley. There, an earthen feature known as the Purron or the Mequitongo Dam (García Cook 1985:30) was constructed across the Arroyo Lencho Diego. Located 2.5 kilometers northwest of the town of San José Tilapa, this dam is only 1 kilometer south and downstream of the Abejas and Purron caves, where the earliest evidence of domesticated plants was found in the New World. According to Woodbury and Neely (1972:43, 84–85), who studied this dam in detail, construction began between 750 and 600 B.C. Initially the dam was much smaller than it would eventually become after three additional periods of construction. After the first building phase, the dam was 175 meters long and had an elliptical cross section with a maximum width of 6 meters and a maximum height of 2.8 meters. It served to impound water, and the resultant reservoir measured 140 by 170 meters. Woodbury and Neely (1972:83–84) claimed that it was built in order "to provide irrigation [water] for their extensive fields, which must have been situated on the large alluvial flats to the southwest [downstream of the dam]." They go on to say, "Although we have no direct evidence of a spillway, it seems probably [sic] that some arrangement of a spillway and canals were present to drain the reservoir and take the water to the fields." The fact remains, however, that confirmed evidence of canals this early anywhere in the Tehuacan Valley was at the time, and still is, lacking (García Cook 1985:22, 30).

There has been much speculation about early irrigation in the Tehuacan Valley. For example, Richard S. MacNeish has argued that irrigation was being employed there between 900 and 200 B.C. He has, however, absolutely no confirmed evidence of canals on which to base such a claim. His conclusion is based on inferences made in large part from late canals and known locational characteristics of early settlements (MacNeish 1962:38). Recognizing the paucity of his own data, MacNeish originally made only cautionary suggestions about early irrigation. Through time, however, he began to consider the interpretations as fact and embellished them considerably, even though there was no additional evidence on which to base his more recent and definitive argument. For instance, in February of one year MacNeish (1964a:536) stated that the Santa María Phase "*may* be the period in which irrigation was first used" (emphasis added). With no more or better data, he claimed unequivocally in November of the same year that the "Santa Maria people of Tehuacan began to grow their hybrid corn in irrigated fields" and "by about 200 B.C. . . . the valley now *had* large irrigation projects" (MacNeish 1964b:10; emphasis again added; see also MacNeish 1971:314–315; 1972:82).

Although MacNeish's claims are widely accepted (e.g., Matheny and Gurr 1983:81), the earlier statements made by Woodbury and Neely, the actual investigators of relic agricultural features in the Tehuacan Valley, indicate that there is no confirmed evidence of early irrigation canals (see also García Cook 1985:22, 30). Indeed, there is no real evidence that Purron Dam actually served an agricultural purpose (García Cook 1985:31). If it did, however, it is more likely that any possible canal was a short spillway extension that facilitated the controlled flooding or watering of fields immediately below the dam (Flannery 1983:333) rather than a device for transporting water to a specific location some distance downstream. Even during the later phases of dam construction, parallel rock alignments claimed to be a canal by Woodbury and Neely (1972:90) ap-

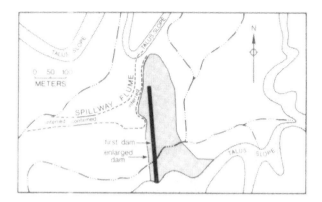

Figure 2.4. Map of Purron Dam showing the spillway and flume. After Woodbury and Neely 1972.

pear more like a spillway apron that would have helped reduce the erosive force of the discharge or overflow below the dam, and a short flume that directed water back into the natural stream channel (Fig. 2.4). In this respect, the dam might have been more of a floodwater protection device for arroyo bottom fields further downstream than a feature associated with canal irrigation.

The first documented evidence of a canal associated with the Purron Dam appeared ca. 600 B.C. (Woodbury and Neely 1972 : 93). This canal did not take water to fields downstream from the dam, however, but rather was located above the dam and apparently carried water to the spillway from further upstream during times of drought when the reservoir was low (Woodbury and Neely 1972 : 90). It was not used to irrigate fields.

Tlaxcala

Although irrigation canals were not used in conjunction with the Purron Dam, as some have long thought, there is tentative evidence that canals were used with storage dams not too far from the Tehuacan Valley in the present-day state of Tlaxcala. According to Angel García Cook (1981 : 245), terraced fields, which had been dry farmed since approximately 1200 B.C. (Abascal and García Cook 1975 : 200) and were similar to those at Chalcatzingo, were being enlarged and improved by the second half of the Texoloc Phase, which extended from 800 to 400/300 B.C. (García Cook 1986 : 255). It is reported that the changes that took place at that time involved the construction of dams across drainages upstream of the terraces. Canals that were

U-shaped and measured from 60 centimeters to 1.1 meters wide at the top and 80 centimeters to 1.3 meters deep were excavated in order to transport water unspecified distances from the newly created reservoirs to distribution canals on the terraces. These branch canals ranged in width from 35 to 60 centimeters, were between 40 and 70 centimeters deep (Abascal and García Cook 1975:203), and facilitated manual irrigation (García Cook 1986:255), or the transfer of water from canals to crops by means of hand-held pots.

Detailed descriptions of the Tlaxcalan dams are lacking. The best available source (Abascal and García Cook 1975:202) says only that they were 4 to 5 meters high, approximately 1.5 meters thick, and built entirely of cut and fitted, rather than uncut and stacked, stone. The reservoirs were also quite small, averaging 10 meters in diameter.

Whether or not the irrigation system at Tlaxcala functioned as reported at the time claimed is debatable. In terms of technology, the dams seem far too great an advancement over either the earthen Purron Dam or even the piled rock dam at Teopantecuanitlan, which itself is dubious. Although they did not impound large volumes of water, these dams appear to be too thin to have supported the load created by a body of water 4 meters deep. It may well be that these dams have other attributes (e.g., very thick bases) that contributed to their strength. Such qualities, however, have not been described.

Also contributing to doubts about these systems are problems with their dates. As stated previously (Chapter 1), their chronology is regarded as suspect by some highly regarded Mesoamerican archaeologists. Accordingly, until additional details about these dams are provided, existing interpretations cannot be accepted categorically. The described networks of canals were, in all likelihood, used at the times claimed. They were, however, certainly no more technologically complex than those at Santa Clara Coatitlan or on the Meseta Poblana.

Monte Albán

The earliest undisputed evidence of a canal taking water out of a reservoir created by the construction of a storage dam is on the east slope of the Xoxocotlan Piedmont, just west of Oaxaca city. Discovered by James A. Neely (Blanton 1978:54), this small irrigation system (Fig. 2.5) is located just below the site of Monte Albán. Remnants indicate that the dam was 10 meters high at the center and 80 meters long. It was V-shaped in plan with the apex pointing upstream (R. Mason et al. 1977:567). Although not rounded in plan, this feature appears to be an early attempt at constructing an "arch dam," one

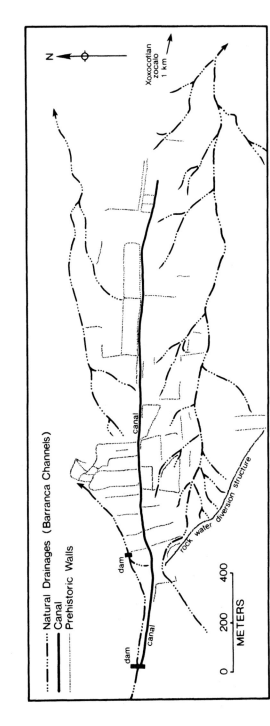

····· Natural Drainages (Barranca Channels)

——— Canal

········· Prehistoric Walls

dam

canal

dam

canal

canal

rock water diversion structure

0   200   400

METERS

Figure 2.5. Map of canal system at Monte Albán Xoxocotlan. After O'Brien et al. 1982.

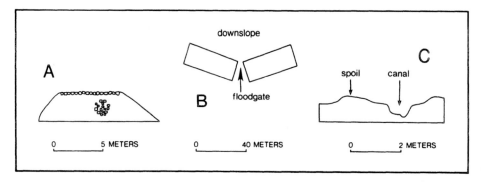

Figure 2.6. Various irrigation features at Monte Albán Xoxocotlan: A, cross section of the dam; B, plan of the dam; C, cross section of the canal. After O'Brien et al. 1980.

that is curved and dependent on arch action for its strength. Such dams are typically thin structures that require less material than gravity dams (N. Smith 1971:263).

The cross-sectional shape of the dam was that of a trapezoid, 12 meters wide at the bottom and 6 meters wide across the top (Fig. 2.6A). The upstream face was notably steeper than that of the downstream side (O'Brien et al. 1980:348). Much like the purported dam at Teopantecuanitlan, the Xoxocotlan dam was constructed mostly of unmodified boulders. This one, however, was cemented by a limestone mix and covered with a thick limestone plaster. Tightly fitted cut limestone blocks formed the uppermost meter. In spite of severe erosion, evidence suggests that a floodgate (Fig. 2.6B) controlling the flow of water into the canal was located in the center of the dam (O'Brien et al. 1980:350). Such devices are not known from earlier sites. Until it can be demonstrated otherwise, floodgates will have to be considered as having their origin here.

It is, of course, possible that floodgates were developed out of technology used for drainage. Indeed, debris that undoubtedly clogged drains such as those at Olmec sites may well have provided irrigators insight into how floodgates operated. In sum, by having a mortared, V-shape masonary construction and a floodgate and being later in date, this dam is considerably more complex technologically than any earlier dams, confirmed or suspected.

The main canal that led from the dam ran approximately 2 kilometers. Through its first 300 meters, the canal follows the general contours of slope until finally reaching the crest of a low ridge. For the remainder of its length, the canal follows the ridge top and provides water to approximately 50 hectares of fields on the gentle slopes on each side. The canal runs off the ridge at its lower end and empties into an arroyo.

Unlike most prehistoric canals, which have either U-shaped, rectangular, or trapezoidal cross sections, this one was stepped (Fig. 2.6C). It is described as having a small "lower canal" 30 centimeters wide and 12 centimeters deep within a larger "upper canal" 80 centimeters wide and 25 centimeters deep (R. Mason et al. 1977:567). The exact purpose of this bi-level canal is unknown. Roger D. Mason and his colleagues (1977:567) think the lower canal might have been used to transport water during the dry season when lesser amounts would have been available, and, hence, would have required close and careful control. There is, however, another equally plausible explanation. It might well be that the upper and lower canals reflect two different phases of construction. If so, the upper canal might have been used at one time to carry small amounts of water to only a few fields during the earliest period of use. Later, when more fields were added, and, hence, more water needed, the canal might have been enlarged by deepening it to its currently known size and shape. Whatever its exact function, the unusual cross-sectional shape does attest to two things—experimentation and an advancement in canal technology. There is, however, another aspect of this canal that is equally, if not more impressive and technologically significant.

In one place along the ridge top the canal is "chiseled into bedrock" (R. Mason et al. 1977:567). Technologically, this is important, not so much because cutting tools were used (although that is important), but rather because it required both planning (foresight) and engineering (advanced skills). Early canals excavated in soft, relatively relief-free alluvium could have been constructed easily by trial and error. That is, builders could have repeatedly gone through a process whereby they sequentially dug a short section of canal, allowed water to enter in order to test its flow characteristics, and then made some adjustments in the canal alignment and gradient before extending the canal by repeating the sequence (see, e.g., Hopkins 1984:77). With an outcrop obstacle on a ridge top, however, they would have been forced to plan a course either around or through it. The builders would have to have known well in advance the entire course of the canal and what was needed to make it work. The people who built this system were, therefore, showing the first true signs of irrigation planning in Mexico.

No evidence was found of branch canals leading from the main canal at Xoxocotlan (O'Brien et al. 1982:33–34). That many fields are located along and adjacent to the canal suggests that whenever water was needed, the floodgate on the dam was opened and water was allowed to flow through the canal. Just as no branch canals were found, there is currently no evidence that openings existed in the canal walls that could have allowed for water to enter the fields (James A. Neely, personal communication, 1987). Exactly how fields were watered, therefore, remains unknown. Presumably, some type of fill, not unlike that found in one of the canals at Santa Clara Coatitlan, could have been placed in the canal at strategic points where needed. Such an obstruction would have blocked the canal and forced water to overflow the canal and flood adjacent fields. This flow was retained on the fields by a series of low terrace walls constructed of both dry-laid and mortared unmodified cobbles and boulders (O'Brien et al. 1982:35).

Additional details of field borders have not been reported. The existence of these features alone, however, is important in terms of technological accomplishments. Terracing technology is probably just as ancient, if not more so, than that of canal irrigation (see Donkin 1979:17). The earliest known terraces were clearly not irrigated, but rather functioned by conserving soil on intensively cultivated slopes in the highlands, where rainfall is not only sufficient for agriculture, but excessive. Although having origins, and undoubtedly an entire sequence of development, distinct from that of irrigation, terracing technology appears to have merged with canal technology for the first time here at Xoxocotlan. This merger of technology ushered in the practice of irrigating bordered checks. In terms of water control techniques, this was a major improvement over wild flooding.

Sherds found in the canal fill indicate that the canal system dates from the Late Formative Period, 550 to 150 B.C. (R. Mason et al. 1977: 567). Although the exact sequence of construction is unknown, canal building might have been initiated between 550 and 400 B.C. with the system being expanded and improved coevally with local population increases until 150 B.C. The canal system appears to have been abandoned at approximately A.D. 250 (R. Mason et al. 1977:572).

One of the real curiosities of this canal involves the water source. Michael J. O'Brien and his colleagues (1982:22) claim thảt the system depended on runoff impounded behind the dam. This explanation seems sufficiently plausible at first glance. Upon closer scrutiny, however, it becomes apparent that further qualification is needed. Put simply, the drainage basin upstream of the dam is approximately 50 hectares in area, or about the same size as the fields that were irrigated. Studies conducted elsewhere indicate that catchment areas fifteen times as large as irrigated areas are barely large enough to supply ample runoff (Tadmor, Shanan, and Evenari 1960). Clearly, the area at Monte Albán was of insufficient size to supply adequate runoff for the Xoxocotlan fields.

It is, however, possible that because Monte Albán was a built-up area, runoff from the site was much greater than it would have been had the hill not been modified by humans. Runoff was certainly a problem for the inhabitants of Monte Albán, as the presence of numerous drains at the site attests (Winter 1985 : 101). It is also possible that human activity itself at the site resulted in artificially increased runoff. Presumably, large quantities of water were carried to the site for domestic purposes. Disposition of waste water would, therefore, have been a problem. In effect, the dam might well have been constructed in order to create a cesspool, the effluent from which would have been used to both irrigate and fertilize the Xoxocotlan fields.

Another possibility is that a spring may once have existed either near or somewhere further upstream from the dam (James A. Neely, personal communication, 1988). Springs certainly exist in natural drainages around Monte Albán (Winter 1985 : 117), and elsewhere at the time in Mesoamerica they were beginning to be exploited in conjunction with dams (e.g., García Cook 1985 : 36).

Development of Spring-fed Systems

The use of springs for irrigation purposes has long been considered a basic agricultural strategy, and one that some scholars (e.g., Steward 1929; Woodbury and Neely 1972 : 128) think may have been responsible for the origins of canal technology. In fact, however, there are relatively few sites in Mexico where springs were tapped for agricultural purposes in prehistoric times, and none so old that it could be considered the birthplace of canal irrigation. Of the known spring-fed canal sites, the earliest is, perhaps not surprisingly, in present-day Oaxaca state.

Hierve el Agua

By far the best-known and best-documented spring-fed irrigation canal complex known from prehistoric Oaxaca, or all of Mexico for that matter, is the site of Hierve el Agua in the mountains east of Oaxaca city near the town of Mitla. Also found by James A. Neely (Flannery and Marcus 1976 : 378), this system involved approximately 50 hectares of artificially terraced hillside (Fig. 2.7). Rows of parallel distribution canals each measuring approximately 30 centi-

Figure 2.7. View of travertine-encrusted spring-fed canals and terraces at Hierve el Agua. From Neely 1967.

meters wide and 30 centimeters deep and with nearly imperceptible gradients carried water across the fronts or tops of rock-faced terraces. These canals were fed by main canals approximately 50 centimeters wide and 50 centimeters deep that began below a cliff on top of which are a series of springs (Fig. 2.8). The system functioned by water from the springs flowing down the cliff face and then being collected by the main canals. These, in turn, carried the water down the slope and into the distribution canals.

The water from these springs was rich in calcium carbonate. As a result, travertine has accumulated on the cliff face and along the canals, thereby fossilizing them over centuries of use (Woodbury and Neely 1972:136). The system appears, by differences in the amount of travertine built up on various canals, to have been expanded through time (James A. Neely, personal communication, 1988). Unfortunately, techniques for dating travertine have not been perfected (Karl W. Butzer, personal communication, 1988). As can best be determined by radiocarbon dating of associated materials, this system came into use ca. 300–400 B.C. and was used through approximately A.D. 1350 (Neely 1967; 1970:85).

The possibility that the Hierve el Agua system was not used for agricultural purposes has been raised. William P. Hewitt, Marcus C. Winter, and David A. Peterson (1987) recently argued that the canals were employed in the production of salt, and Winter (1985:119) himself once thought that they might also have functioned to feed mineral baths. These are certainly interesting and provocative interpretations. They are, however, wrong for at least two reasons. First, although they claim that features similar to some of those found at Hierve el Agua were used for salt making at the not-too-distant Fábrica San José site, the features of the respective sites are actually quite different in terms of both morphology and chronology (Doolittle, in press). Indeed, there is no comparable evidence of salt production as described by Hewitt, Winter, and Peterson for any Meso-

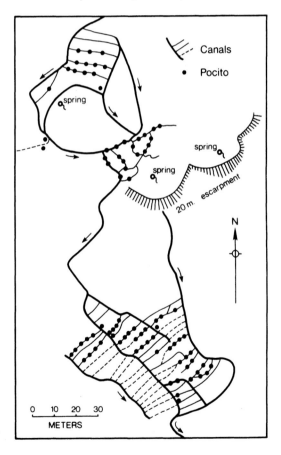

Figure 2.8. Map of canals at Hierve el Agua. After Kirkby 1973.

american site as old as Hierve el Agua. Second, although they claim that the spring water at Hierve el Agua is unsuitable for crop production and that, if it were suitable, agriculture would be practiced there today, Neely (1970:85) pointed out that the water is suitable for traditional though not modern hybrid varieties of crops if used carefully, and that the terraces were cultivated until 1966, when local officials put an end to the practice after they were informed of the archaeological significance of the site. In sum, most archaeologists accept that the canals and terraces were used for the irrigation of crop plants (e.g., Flannery 1983:327), and there is no conclusive evidence that they were not, especially in their earlier years.

Because this system is younger than any of those discussed previously, it can be concluded that there was no way that canal irrigation could have originated from spring diversion, or that the ancient peoples at Hierve el Agua developed a canal system independently. Instead, it is most likely that the earliest farmers at Hierve el Agua initially planted crops in naturally watered places below the cliff, and later adopted canal technology developed elsewhere in order to expand cultivation to include places that were not watered naturally.

Being the earliest confirmed spring-fed canal system in Mexico, Hierve el Agua manifests one significant technological accomplishment. Unlike earlier systems that involved the diversion of periodic floodwaters or the release of water from a storage dam built across an ephemeral stream channel, this one required that irrigators deal with permanently flowing water. That was no easy task, but one that was apparently managed by having canals that carried water past and through fields, much as in the case of the main canal at Monte Albán Xoxocotlan, rather than to or terminating in them as in most other floodwater farming systems. Getting water out of the canals then became the greatest problem, and this was apparently handled in one of two possible ways.

Many researchers (e.g., Kirkby 1973:117) think that the ancient farmers at Hierve el Agua practiced pot irrigation, removing water from the canals by means of hand-held vessels. As evidence for this interpretation, they point to circular depressions, known as *pocitos*, that are found within many of the distribution canals and spaced at roughly regular intervals of 3.25 meters (Figs. 2.8 and 2.9). Although the exact function of these features remains unknown, it has been argued that they were constructed in order to accommodate dippers that otherwise would not fit into the small canals. This argument seems logical enough and is supported by ethnographic parallels. In parts of Mesoamerica today, and particularly in Oaxaca, farmers have been seen irrigating small plots by removing water manually

Figure 2.9. View of distribution canal with *pocitos* on the top or front edge of a terrace wall. Note the manner in which this canal takes water out of, or away from, the rather steep gradient main canal in the rear. Courtesy James A. Neely.

from widened points in otherwise narrow canals (Wilken 1987: 182–193, especially p. 183 and Fig. 9-16).

A second scenario for the removal of water from the canals has been proffered by Neely (personal communication, 1988), but, unfortunately, does not appear in print. According to him, whenever water was needed on a particular terrace, earthen materials were piled into the distribution canal near its downstream end, thereby blocking the flow of water. Because each distribution canal had a nearly imperceptible gradient, water would have backed up and spilled over the side of the canal along its entire length, thereby sheet flooding the field on the terrace below. The flow of irrigation water across the field could have been stopped by removing the fill in the canal and allowing water to discharge through the network. Field flooding could also have been stopped by placing earthen fill in the distribution canal at its upstream end, near where it bifurcated the main canal.

The blocking of distribution canals at the points where they intersect the main canals was undoubtedly a common practice at Hierve el Agua regardless of the means by which fields were watered. The discharge from the springs was never so great that water flowed abundantly and constantly through the network of canals. This was especially true in drought years, when the discharge from the springs would have been abnormally low. In order to compensate for such deficiencies, irrigators would have had to take turns using the scarce hydraulic resource. As a result, the canals at Hierve el Agua were the first that involved what could be called true sluice gates. This was a major technological development. It combined floodgate technology and canal blocking—both employed at Monte Albán—with branch canal technology used at places such as Meseta Poblana. It is a particularly important technological breakthrough in that it allowed for control of permanently flowing water and, therefore, led the way for tapping other, larger perennial sources.

Other Prospects

In addition to the relatively few studies of ancient irrigation that are based on direct evidence of canals, there are several more for the state of Oaxaca that are not. Indeed, admissions of the lack of direct canal data in that area are legion: "traces of early irrigation canals are virtually eradicated" (Flannery et al. 1967 : 451); "there is no evidence that canal irrigation was practiced" (Kirkby 1973 : 127); "So far no ancient canals have been detected" (Lees 1973 : 90); "No canals [were] definitely identifiable"; "we located no clear traces of Late Formative canals" (Whalen 1981 : 87, 103); "We found no evidence of sophisticated canal irrigation technology" (C. S. Spencer 1982 : 82); "there is no direct evidence in the form of Precolumbian canals" (Drennan and Flannery 1983 : 65); "We have found no evidence for canal irrigation facilities" (Redmond and Spencer 1983 : 119). For the most part statements about canal irrigation during this phase have been made on the basis of environmental conditions and possibilities, present-day analogs, and the locations of known ancient settlements (Winter 1985 : 104, 109, 113–114).

The consensus of opinion among scholars working in both the

Valley of Oaxaca and the Cuicatlan Cañada of Oaxaca is that the earliest farmers lived on the high alluvium along the floors of the large
valleys. The river that formed these valleys did not have a constant
surface flow because much of the water from the perennial tributaries was absorbed by the coarse alluvial sediments of the floodplain. As a result of the unusually high water table there, early farming is thought to have relied principally on pot irrigation with water
taken from shallow wells on the lower, cultivated alluvium (Flannery et al. 1967:449). It has also been suspected that, in places, water
may have been diverted from streams by means of stake and brush
weirs (e.g., Redmond 1983:63) and possibly carried to the fields in
"small gravity-flow canals" (Flannery and Marcus 1976:378; see
also C. S. Spencer 1982:82).

Conclusions about canals are, of course, not based on direct evidence, but rather inferred from the locations of settlements (e.g.,
Blanton and Kowalewski 1976). Furthermore, that canals were used
seems most unlikely. In her detailed and highly regarded (see Butzer
1982:155) analysis of Oaxaca agricultural environs, Anne V. T.
Kirkby (1973:128) argued that areas near streams amenable to diversion were watered by using the "simplest floodwater techniques."
She goes on to say, however, that although "canal irrigation as opposed to simple stream diversion was not practiced extensively . . .
experiments with irrigation may well have been initiated" prior to
300 B.C. Here, Kirkby distinguishes floodwater diversion from irrigation on the basis that the latter involves canals and the former does
not. If such irrigation was employed, however, it must have been
small in scale, as there are no suitable large or expansive areas in
either the Valley of Oaxaca (Flannery et al. 1967:449) or the Cuicatlan Cañada (E. Hunt 1972:195). Furthermore, such rudimentary
canals would not have been much more advanced than the system
documented earlier at either Santa Clara Coatitlan or the Meseta
Poblana.

There also have been suggestions that canals were beginning to be
used in the upper piedmont areas, because sites began to be situated
in proximity to perennial tributaries at this time. Many sites dating
to this period are located along these streams, "not downstream, at
the point where most water is available, but upstream, where water
can be diverted for irrigation" (Flannery et al. 1967:451). Two such
sites are Santo Domingo Tomaltepec, in the Valley of Oaxaca, and
Cuicatlan, where there exists confirmed evidence of both later prehistoric (Whalen 1981:21; Hopkins 1984:100–105) and present-
day (Lees 1973:92; Hopkins 1983:269) irrigation canals (Fig. 2.10).

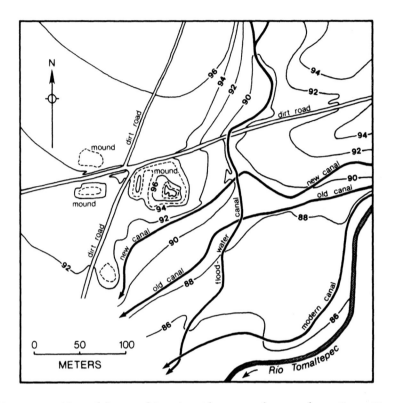

Figure 2.10. Map of late prehistoric and present-day canals at Santo Domingo Tomaltepec proffered by some as evidence to suggest that earlier canals must have been used. After Lees 1973.

Whether or not canals were used for transporting water from perennial streams to fields prior to 300 B.C., as Michael E. Whalen (1981: 87) suggests, remains unresolved. Kirkby (1973:119) noted that "canal irrigation, as practiced today, does not depend on anything more than the simplest technology." On the basis of this analog she then suggested that "There is no technological . . . reason why present day canal irrigation schemes could not have been operated in the Formative period."

Kent V. Flannery (1970:79), however, was more cautious when he said that "future work is certainly warranted—perhaps even an attempt to trench for Late Formative period irrigation canals on the slope below the site." As was the case with the Xoxocotlan canal, "irrigation systems associated with the hilltop centers must be found" (O'Brien et al. 1982:22). Until such work is conducted and reveals otherwise, we can conclude only that the diversion of con-

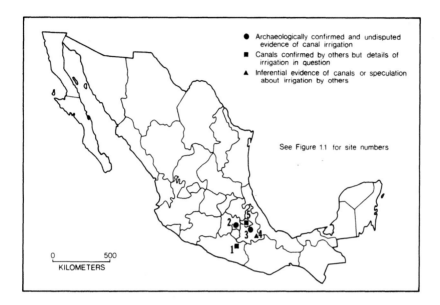

Figure 2.11. Locations of canal irrigation, 1200–600 B.C.

stantly and rapidly flowing streams did not occur until after 300 B.C. The data indicate that these piedmont areas were not only known, but also utilized, "long before their systematic development by irrigation farmers" (Flannery 1970:80).

Although confirmed, undisputed data are minimal, evidence indicates that canal irrigation underwent small, but marked and important technological developments between 1200 and 300 B.C. The only canals used to transport water to agricultural fields prior to 600 B.C. were those that involved diversion from ephemeral streams. There exists some tentative evidence of canals in valleys immediately to the south of the volcanic axis that crosses Mexico from east to west. Confirmed evidence of early floodwater irrigation, however, comes only from the valleys in the southern part of the Mesa Central (Fig. 2.11).

Between 600 and 350 B.C. the practice of canal irrigation appeared further to the south in what is today the state of Oaxaca (Fig. 2.12). Confirmed evidence is found at only two sites there, but these data illustrate that new technologies were being developed. Different water sources were being used, and these involved new problems that had to be overcome. Substantiated evidence of spring-fed canals

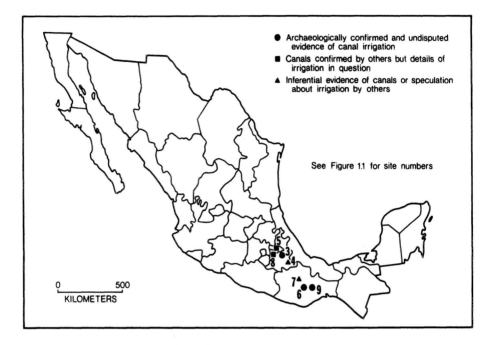

Figure 2.12. Locations of canal irrigation, 600–350 B.C.

exists from only one site. There is, however, a great deal of speculation that permanent water sources, particularly mountain streams,
were beginning to be used for irrigation ca. 300 B.C. The locations of
sites with such evidence are found in a chain of valleys—Oaxaca,
Cuicatlan, Tehuacan, and Puebla—that could be considered the
backbone of the northern part of Mesoamerica.

3. A Time of Maturation, 350 B.C.–A.D. 800

Incipient Use of Upland Streams

The diversion of water from perennial or constantly flowing streams for irrigation purposes began, in all probability, in what is today the state of Oaxaca. There, the high alluvium of the valleys was brought under irrigation, but not by diverting water from the main river channels as might be expected. Instead, water was diverted from the tributary streams (Whitecotton 1977:33).

Loma de la Coyotera

Albeit scant, the earliest confirmed evidence for such diversion came from the Cuicatlan Cañada, and specifically from the Loma de la Coyotera site. There, canals have been dated to the Lomas Phase, which began ca. 300 B.C. (Redmond 1983:123). This site is located on a ridge between two streams that head in the sierras where rainfall is abundant. Although discharge is not great, averaging only about 0.07 cubic meters per second (Kirkby 1973:46), the velocity is high because the streams are deeply incised and their gradients are quite steep. In spite of these apparent difficult environmental conditions, however, irrigators today, and presumably in prehistoric times as well, have little difficulty in transporting irrigation water through canals because the source is considerably higher in elevation than the fields (E. Hunt 1972:195).

The exact nature of the devices used prehistorically to direct water from the streams and into canals is unknown, as they have been long destroyed. Today, however, diversion dams built principally of stakes and brush, but anchored by stone and stabilized with earth, are commonly used (Hopkins 1984:75; see also Kirkby 1973:119 and Flannery 1983:328). Because of their apparent simplicity, it

seems likely that similar features would have been used in prehistoric times (Winter 1985 : 104, 109).

Although such features are not very elaborate, they actually constitute an important step in the development of irrigation technology. Unlike weirs, such as those proposed as having simply diverted
ephemeral flood waters onto the lower alluvium of the valley bottoms, and unlike storage dams that impound water, diversion dams
raise the level of the water in order to allow it to flow into canals
whose floors are higher than the otherwise normal surface of the
stream flow. They reflect, therefore, the marriage or combination
of two types of devices, each with separate and distinct functions,
into one.

Because they are totally submerged and subjected constantly to
the force of moving water, diversion dams made principally of brush
require regular maintenance. They also need constant attention in
order to control the amount of water flowing into the main canal. To
this end, head gates involving earth and rock fill that could be inserted to keep water out of or removed to let water into the canals
were probably used. Such devices were simply the adaptation of
sluice gate technology developed to regulate the flow of water between main and distribution canals at Hierve el Agua. Presumably,
maintenance on diversion dams could have been done as part of
this regular water control activity. Whatever the particular details
may have been, however, diverting water out of constantly flowing
streams required more labor, control expertise, and technology than
either diverting periodic flood waters or releasing water from a dam
across an ephemeral stream channel.

The main canal at Loma de la Coyotera is not very long, probably
owing to the steep gradients of the tributary stream from which
water was taken. It was traced for approximately 1,400 meters (C. S.
Spencer 1982 : 224), but appears not to have been more than 2 kilometers long overall. Toward its downstream end, this canal split into
a series of four small distribution canals (C. S. Spencer 1982 : 225).
Details about these canals, including their sizes, have not been reported, but some type of sluice gates must have been used, and
Charles S. Spencer (1982 : 222) suggests that as much as 737 hectares
of high alluvium could have been watered. It is entirely feasible that
such a great amount of land could have been irrigated later in prehistoric times as other canals were added. It is unlikely, however,
that so much land was irrigated so early on. In all likelihood, only a
small percentage of the land that eventually was irrigated received
water via canals when the system was first pressed into service. If

10 percent is accepted as a conservative estimate, then approximately 75 hectares of the land could have been irrigated when the system was first built.

Although the length of the main canal and the branching toward its downstream end are not unusual, there are two things that make this canal both interesting and important from the perspective of being a technological advancement. First, although it had a rather steep gradient (dropping 35 meters over its traceable length, for an average gradient of approximately 2.5 percent), it does not seem to have been affected by erosion to the same extent as the canal found at Santa Clara Coatitlan. Presumably, this can be attributed to better control of water flowing into the canal from a permanent as opposed to an ephemeral source. Second, aqueducts were found in twelve places where the canal crossed small barrancas or narrow channels of ephemeral streams that feed the main tributary stream. For the most part, these were little more than earthen fill across the extreme upstream ends of very small gullies. In one place, however, an aqueduct measuring 60 meters long, 4 meters wide at the base, and 1 meter high was built (C. S. Spencer 1982:224).

The construction of such features, much like the excavation in rock noted for the Xoxocotlan canal at Monte Albán, shows both prior planning and engineering; the builders knew well ahead of construction what they were doing and how to do it. Unlike removing or excavating a short section of material, however, the filling-in of an area, especially one 60 meters across, is difficult and involves advanced understanding (Zimmerman 1966:342–344). The removal of material is relatively simple as compared to filling because problems such as the suitability of building materials and how to handle runoff flowing through the channel being crossed are not a factor. Furthermore, the sheer idea of transporting water above the natural ground level to some distant point is in itself an abstract concept that inexperienced irrigators probably could not have conceived. The builders of the canal system at Loma de la Coyotera, therefore, knew what they were going to do well before construction actually took place.

Canal systems similar to the one just described have been proposed as having existed in the Valley of Oaxaca during this period (Flannery et al. 1967:451; Kirkby 1973:133, 135; Lees 1973:90, 92). Settlements dating as early as 300 B.C. are located in similar settings. Although no concrete evidence of canals, other than modern or present-day ones, has been uncovered, there should be little doubt that such systems did indeed exist.

Cuicuilco

Other evidence for the use of perennial streams for irrigation in Mexico at this time is sketchy, in spite of the relatively large amount of archaeological work that has been done. One place that has evidence of canals that may have tapped a constantly flowing stream is the famed circular pyramid site of Cuicuilco in the southern part of modern Mexico City. There, in 1956, Angel Palerm (1961b: 300), in the company of Eric R. Wolf, found remains of two irrigation canals. Details of these *"zanjas"* or *"canales"* are unknown for three reasons. First, with the exception of a few places, they are covered by an extensive lava flow. Second, other than the limited amount of work that was conducted immediately after their discovery, there has been no systematic study of these canals (Wolf 1959: 77). Third, as throughout the Basin of Mexico, ever-expanding Mexico City has wiped out much of the potential evidence. What is known is that the canals ran down a hillside, but from an unknown or, at least, undescribed source. They might well have originated at a small perennial tributary that was fed by springs.

The significance of the Cuicuilco canals lies not so much with their source of water, however, as it does with their age. Until recently, these canals were considered to be the oldest in the Basin of Mexico. William T. Sanders (1965: 16, 114–115, 171) even felt that the concept of irrigation diffused from Cuicuilco to the Teotihuacan Valley and then throughout the basin (Sanders 1981: 170–192). The Cuicuilco canals were initially thought to have been constructed prior to 500 B.C. (Palerm 1961b: 300). Later, however, Palerm (1973) accepted the dates from earlier pyramid excavations (Heizer and Bennyhoff 1958: 232–233) and agreed that the canals were probably dug ca. 300 B.C. These findings now seem to be commonly accepted (e.g., Sanders, Parsons, and Santley 1979: 273). Although it is now known that the canals at Santa Clara Coatitlan (Nichols 1982a) date much earlier, the ones at Cuicuilco remain the first known to have been used in the Basin of Mexico in over four hundred years.

Early Diversity in the Central Basins

Beginning at approximately 300 B.C. the use of canals for irrigation virtually exploded throughout the central highland basin (Figs. 1.1, 3.1). The exact nature of this expansion and the technological developments that occurred with it are not well understood. Sanders (1976: 116, 124), however, once maintained that the earliest post-

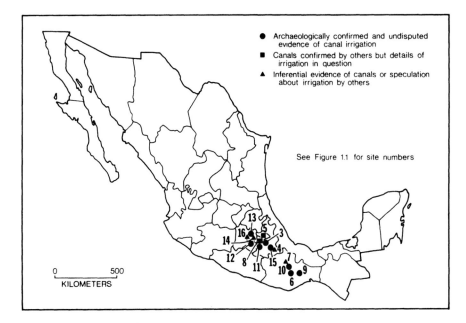

Figure 3.1. Locations of canal irrigation, 350 B.C.–A.D. 200.

Cuicuilco irrigation involved the use of permanent, constantly flowing spring-fed streams in the Teotihuacan and Texcoco areas. This claim was probably based on his understanding of the then earliest-known canals at Cuicuilco and the then available settlement and environmental data from the northeastern and eastern parts of the Basin of Mexico.

Amanalco

Near Amanalco, Texcoco, south of Teotihuacan on the east side of the Basin of Mexico, the drainage pattern consisted of small, shallow spring-fed streams during the period in question. According to Sanders (1976:124; Sanders, Parsons, and Santley 1979:270), communities situated along these channels would have been able to divert stream flow into simple irrigation canals and then onto their fields only a few hundred meters away.

Given the paucity of data, the alleged Amanalco system purported by Sanders was probably no longer nor more complex than any other earlier or contemporaneous canal system in Mesoamerica. Virtually all of the components needed in order for the system to function as

Sanders says had been used somewhere in the region by 300 B.C. This suspected system might well provide information about the expansion of canal irrigation throughout the Basin of Mexico. It does not, however, contribute much to our understanding of the development of canal technology, and it did not play a significant role in the development of the technology itself.

Otumba I

In contrast to the notion that canal irrigation in the Teotihuacan and Texcoco areas developed from the use of permanent streams is the idea that floodwater irrigation was the predecessor. Floodwater systems are as widespread as permanent systems in the valley (Lameiras 1974:178–179, 191–193), and they have a long history of use (Sanders 1976:103–104, 114–115). Their antiquity, however, was not fully recognized until after a 1977 field reconnaissance by Thomas H. Charlton on the valley floor northeast of Teotihuacan, 500 meters south of the town of Otumba (Fig. 3.2), revealed a series of ancient irrigation canals, exposed in a road cut (Charlton 1977). Detailed excavations on both sides of the road during the summers of 1977 and 1978 uncovered several canals. Assessment of the surrounding physical environs indicated quite clearly that the canals had as their source of water the ephemeral, or only seasonally flowing, Barranca del Muerto.

Charlton's (1978:33) analysis of the stratigraphy, as well as his observations of floodwater farming in the area today, revealed that sedimentation due to excessive flooding was a common hazard to ancient irrigators at Otumba. Rocks washed into the canals tended to be numerous and water-worn, indicating that substantial amounts of rapidly flowing water, running off and eroding upland slopes, entered the canals periodically (Charlton 1978:32–33). This sedimentation had resulted in the canals being re-excavated and realigned by their users a number of times.

Given the differences in elevations relative to their spatial proximity, as well as variations in associated ceramics, it is clear that the canals could not have constituted parts of one system. Instead, they were components of several different systems, used at different times (Charlton 1978:31–32, 19–21). The earliest canal dates to the Terminal Formative. Charlton (1979a:11) concluded that it was used in Patlachique/Tzacualli times, ca. 100 B.C. Working on the assumption that the canals were in use at that time, William T. Sanders, Jeffrey R. Parsons, and Robert S. Santley (1979:267–268) concluded

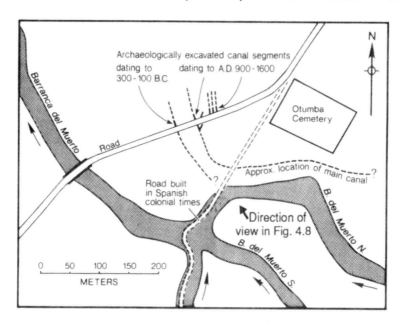

Figure 3.2. Map of the Otumba site. After 1 : 50,000 topographic map, S.P.P. Texcoco E14B21, and Charlton 1979a.

that they had to have been built earlier, perhaps as early as 300 B.C. Perhaps not surprisingly, this early canal was V-shaped in cross section and small, measuring approximately 1 meter wide at the top and 1 meter deep.

The early Otumba canal probably carried water to no more than a few hectares, but it is important for understanding the spread of irrigation throughout the Basin of Mexico. Details about the canal's morphology indicate, first, that it was a technological improvement over the much earlier canals at Santa Clara Coatitlan. Indeed, in terms of the water source it taps, it appears to be comparable to the canals found on the Meseta Poblana. Second, the confirmation of a floodwater diversion system indicates that there was no single type of irrigation system used in the Basin of Mexico at the time. There was, instead, a variety of systems, each developed in its own context—by the borrowing of ideas and technology developed elsewhere—in order to utilize varied local environmental conditions. The canal systems either known or suspected of having been used "suggest significant experience in land manipulation for agricultural purposes" (Matheny and Gurr 1983:81). They do not, however, reflect any advancements in canal irrigation technology. Instead, they seem to indicate a simple expansion of existing technology into areas where it previously had not been used (García Cook 1985:61). According to B. L. Turner II (1983:28), "the number and areal magnitude of irrigation features are [large] . . . for the Basin of Mexico . . . , reflecting to a large degree the greater need for that technology in the more arid basin."

Not only was canal irrigation beginning to be expanded throughout the Basin of Mexico at this time, but advancements were also being made in other central highland basins where canals had long been known. Most notable in this respect is the greater Puebla region.

San Buenaventura

Working in the 1960s, a group of German prehistorians found evidence of ancient irrigated fields approximately 15 kilometers west of the present-day city of Cholula, between the towns of San Nicolás de los Ranchos and San Buenaventura Nealtican. The fields were located on the floodplain between the ephemeral Barranca Tlaltorre and the permanent Río Nexapa—which nearly parallel each other—1.5 kilometers upstream of their confluence. They were covered, and thus preserved, by layers of volcanic materials, principally pumice and tuff, 2.5 meters thick (Seele 1973:77, 78). According to Enno Seele (1973:81, 82), who was in charge of the investigations of the site, the fields were apparently used continuously for over a millennium. Ceramics found in the fields suggest that they were brought into use during the middle of the Late Formative Period, ca. 350 to 150 B.C. Radiocarbon-dated evidence of crops and the lack of eruptions that would have disrupted cultivation indicate that agriculture was practiced until approximately A.D. 1150. The volcanic materials capping the fields were deposited after eruptions of Volcán Popocatepetl in the middle of the twelfth and fourteenth centuries A.D.

The fields at San Buenaventura were discovered as the overlying rock was quarried for use in the construction of buildings in nearby towns. An area measuring approximately 90 by 30 meters of ancient field surface measuring approximately 50 hectares (Seele 1973: Mapas 2, 3) was eventually exposed. Many remnant field features remained intact after exposure. These consisted mainly of parallel ridges and furrows that were probably formed as the result of hoeing. A typical practice throughout much of Mexico today, and presumably in prehistoric times as well, is for farmers to hoe earth into either mounds or ridges around maize plants (Wilken 1987 : 135 – 140). Such earthworks serve to support crop plants against heavy winds, and the resulting furrows both channel and retain irrigation waters. The ridges uncovered near San Buenaventura vary between 1.0 and 1.2 meters apart, and from 20 to 30 centimeters high (Seele 1973 : 80). Most run perpendicular to the direction in which the main streams flow. In a few places, however, some ridges and furrows run at 90-degree angles to others, paralleling the streams (Kern 1973 : 73). Also, in some places, furrows were blocked with earth to the same height as the ridges.

The exact manner in which this field was irrigated remains unknown, largely because unequivocal evidence of main and distribution canals has yet to be found (Seele 1973 : 82). It is reasonably certain, however, that water was diverted out of the ephemeral barranca, not the perennial river. Although specific figures have not been reported, floodplain elevations are highest along the barranca. The furrows follow a rather steep 3.0 percent gradient before terminating near the river. The cross-ridges that subdivided the furrows, therefore, probably functioned to slow the velocity of water, distribute it evenly over the field, and impound it in specific places, thereby allowing it to soak in (Seele 1973 : 80). In effect, they functioned in a similar fashion to the obstructions reported in a field canal at Santa Clara Coatitlan (Sanders and Santley 1977 : 586).

In all probability, the furrows were fed with water from a canal, either long obliterated or yet to be discovered, that had its head somewhere further upstream in the barranca. If so, the canals involved in this system were not much, if any, of a technological development over systems reported on earlier (e.g., the one at Meseta Poblana). What makes this system important, however, is the degree to which water was controlled on the field itself.

Unlike previously discussed older systems, this one did not rely on wild flooding or bunds and bordered checks. Instead, it involved a much more highly developed form of water control. Given that furrows and ridges are, in effect, a dense grouping of field canals and bunds, the fields at San Buenaventura represent the next stage in the development, or the next generation, of water control technology as it pertains to field surface features. Just as bunding, such as that at Monte Albán Xoxocotlan, was a greater technological achievement than wild flooding, furrow irrigation was a development over bordered checks. As with any technological development, of course, ridges and furrows required greater knowledge, labor inputs, and organization on the part of the builders and users. They also helped pave the way for yet greater achievements.

Valley Bottom Developments

In addition to the small-scale field features, some large-scale developments were taking place in the Puebla Valley, and the neighboring Tehuacan Valley as well—an area where evidence of earlier canals has not been found, but has long been suspected. Significant in this regard are the accomplishments made in the irrigation of broad valley bottom floodplains.

Amalucan

South of Tlaxcala in the Puebla Valley is evidence of a water control system that has been dated at 200 B.C., and possibly as early as 500 B.C. On the lands of the Amalucan hacienda, just east of the city of Puebla, Melvin L. Fowler (1969) found, by means of aerial photographs, a series of canal-like features (Fig. 3.3) whose function has recently been determined after several seasons of field investigations. Initial excavations of only one suspected canal revealed that a linear feature approximately 5 meters wide, 2 meters deep, and 1,100 meters long had indeed been excavated in prehistoric times (Fowler 1969:211, 212). That this "canal" carried water from an ephemeral stream channel into suspected branch canals was not confirmed, as none of the latter were tested archaeologically in the early studies. Given a number of conditions, not the least of which were adequate rainfall and a number of mounds and pyramids found on what would have been fields, Fowler (1969:214, 215) originally doubted "that irrigation was needed for agriculture," stating that "this canal system . . . may have been utilized to distribute water to the residents of what must have been a considerable town."

In spite of his initial claims to the contrary, many archaeologists who specialize in Mesoamerica (e.g., García Cook 1985:40; William T. Sanders, personal communication, 1987), as well as many who specialize in other parts of the New World but are concerned with agriculture (e.g., Haury 1976:150) have accepted that the Amalucan canals were intended, at least in part, for irrigation. Although there was at the time little evidence to support this notion, scholars who felt that the canals were not used strictly for domestic purposes may well have been correct.

Later studies by Fowler (1987), involving a detailed surface reconnaissance of the site and its surrounding environs as well as additional excavations, revealed that some branch canals did lead water away from the previously documented main canal at right angles,

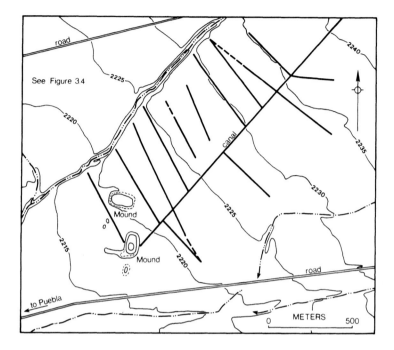

Figure 3.3. Map of canals at Amalucan, Puebla. After Fowler 1987.

while ditches carried water into a channelized stream bed running parallel to the main canal (Fig. 3.4). No details on the morphology of either the branch canals or the ditches have been reported, but Fowler (1987 : 63) says that they differed. The canals that took water from the main canal were shallower than the ditches that carried it into the channelized stream. There were, allegedly, some yet smaller lateral canals, but these too have not been sufficiently described. The only additional information available on the morphology of the canal system is that the main canal and the channelized stream that it paralleled were approximately 380 meters apart and the distribution canals and drainage ditches were 40 meters from each other. The entire system covered only about 70 hectares (Fowler 1987 : 62).

On the basis of his more recent investigations, Fowler (1987 : 61) now feels that the system evolved over a five-hundred-year period from one that was intended entirely for field drainage to one that involved drainage and irrigation. He argues, essentially, that agriculture could have been practiced successfully in the area during most of the year without irrigation, and that the greatest problem for cultivation was poor drainage of the lacustrine soils (Fowler 1987 : 54,

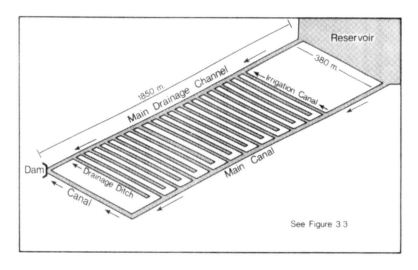

Figure 3.4. Schematic of canal system at Amalucan, Puebla. After Fowler 1987.

53). Once fields were drained, perhaps around 500 B.C., large earthen storage dams were constructed at both the upstream and downstream ends of the canals in order to control the flow of water (Fowler 1987:57). Impounded water, it is postulated, was released during low-rainfall winter months and, therefore, facilitated the cultivation of more than one crop each year (Fowler 1987:60).

One could make the argument that the Amalucan canals did not constitute a canal irrigation system per se, as they did not function primarily to carry water from a source area to fields. Fowler 1987:60) himself suggests this when he says, "The system discovered at Amalucan is very similar to one that Wilken (1969) described."

In his study of present-day cultivation practices and technology not too far from Amalucan, but across the modern Puebla-Tlaxcala state boundary, Gene C. Wilken (1969:234) found that "There appears to be no functional difference between chinampas in the Basin of Mexico and drained fields in Tlaxcala." If chinampas are no different than the fields Wilken studied, and if these are similar to the ones at Amalucan, then it can only be concluded that the latter system reported by Fowler is essentially identical to raised fields reported elsewhere in ancient Mexico (e.g., Coe 1964; Armillas 1971; Turner 1980; Siemens 1983). "Canals" in such systems are low gradient features whose purpose was more to *hold* water than to *trans-*

Figure 3.5. View of the Amalucan hacienda lands from the north. The area under cultivation was that area irrigated prehistorically. The large mound is near the southwesternmost corner of the system discussed by Fowler.

port it. They did not carry water from a source to an otherwise moisture-deficient area so much as they allowed for water that was almost always standing in the canals to be lifted, presumably in hand-held vessels, onto the fields occasionally (Wilken 1987 : 82 – 85). The Amalucan system would certainly then have been a water control system that involved a great deal of construction, maintenance, and operational skills and technology. It would not, however, have been a canal irrigation system in the truest sense of the term.

Regardless of such academic arguments, the Amalucan system was a most important accomplishment that facilitated the development of canal irrigation technology in Mexico during prehistoric times. The Amalucan system is the earliest recorded attempt by people to manipulate the natural hydrology for agricultural purposes on the bottom of a broad, seasonally arid valley (Fig. 3.5). Earlier systems, regardless of their complexity (e.g., that at Monte Albán Xoxocotlan), functional and evolutionary similarities (e.g., that in Tlaxcala), and small-valley-bottom location (e.g., that at Santa Clara Coatitlan), involved the use of relatively steeper gradient canals for transporting water from upland sources. The Amalucan system involved only valley-bottom water and fields.

Llano de la Taza

The idea of irrigating valley bottoms first appeared at Amalucan. From there, however, it appears to have diffused southward quite quickly into the Tehuacan Valley, where it was combined with canal technology that had developed in and diffused northward out of Oaxaca. The result of this blending of two ideas and technologies resulted not only in the first documented use of canals in the Tehuacan Valley but also in the development of a truly remarkable ancient canal irrigation system.

Throughout the valley floor from just north of the city of Tehuacan to south near the town of Axuxco, an area known as the Llano de la Taza, there exist six extensive networks of ancient irrigation canals that remain most visible on the landscape today (Fig. 3.6). Angel García Cook (1985:54) claimed that these systems encompassed over 250 square kilometers. Analysis of topographic maps, however, indicated that he was probably referring to the entire valley bottom. Indeed, this same cartographic assessment revealed that each of the six systems irrigated between 75 and 80 square kilometers or approximately 1,300 hectares. This figure, however, is probably accurate only for late prehistoric times, when the network of canals

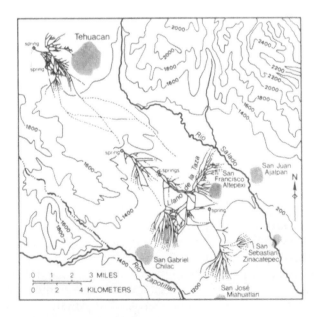

Figure 3.6. Map of fossilized canals on the Llano de la Taza in the Tehuacan Valley. After Woodbury and Neely 1972.

Figure 3.7. Fossilized canal cut by a road on the Llano de la Taza. From Woodbury and Neely 1972.

reached its maximum extent. During the earliest stage of their existence, these canal systems were undoubtedly much smaller. If, as in the case of Loma de la Coyotera, only about 10 percent of the area was irrigated early on, then each network initially involved only about 130 hectares. Although this is not a large area, it is nearly twice as great as any of the earlier known systems.

The sources of water for the Llano de la Taza canals were springs rich in calcium carbonate (Brunet 1967). As a result, they have been "fossilized" by the buildup of travertine. In many places the canals stand as much as 2 meters above the present surface of the land (Fig. 3.7). Soil erosion is in part responsible for this phenomenon (C. E. Smith 1965 : 73–75); another factor, however, is mineral accumulation. Thickness of the travertine tends to suggest that formation either was quite rapid or continued for a long period of time. Woodbury and Neely (1972 : 128) concluded that the latter explanation is most probable because almost all of the material is heavy, hard, and very compact.

On the basis of travertine accumulation and the location of canals to datable settlements, it has been determined that the earliest use of these spring-fed canals must have been sometime around 200 B.C. (C. E. Smith 1965 : 94). Other evidence, such as ceramics embedded in the travertine, indicates that the system was both continuously used and expanded until the time of Spanish contact (Woodbury and Neely 1972 : 128, 135–136). A few such canals were even constructed during historic times (Gil and Neely 1967).

The raising of the canals above the ground surface, although in large part a natural process, was important in terms of the development of canal irrigation technology. Indeed, in concert with rudimentary aqueduct technology developed at Loma de la Coyotera, it provided valuable knowledge that facilitated the irrigation of broad valley bottoms, and it led to improvements in water control.

Earlier canals in Mexico were for the most part constructed through excavation. They were below ground level and terminated at fields that were wild flooded (Fig. 3.8A). The technology involved was minimal and, hence, only small areas could be irrigated. In order to cultivate larger areas, better water control was needed. This could only have been accomplished by extending the main canal farther downstream and building additional water control features (e.g., bunds, sluice gates).

Extending irrigation canals downstream can be done in two ways— through additional excavation or by elevation (Fig. 3.8B). The former method involves not only the excavation for the extension of the main canal, but also the digging of a series of branch canals. The reason for these latter canals involves gravity flow (Zimmerman 1966 : 203–234). The level of the water in an excavated main canal is lower than the surface level of the adjacent field. The only way water can be taken out of the main canal by means of gravity flow is through a branch canal that bifurcates the main canal at a point farther upstream where the elevation of the canal is higher than that of the field. Branch canals can, however, be eliminated if the main canal is elevated. With the canal raised above the ground surface, the adjacent fields can be watered by simply blocking the canal and allowing water to overflow onto them, much as was the case at Hierve el Agua, or by opening sluice gates in the sides of the canal in selected places.

The travertine-encrusted system of the Llano de la Taza involved both elevating the canals and the use of sluice gates. Toward their downstream ends, canals were small, and U-shaped in cross section.

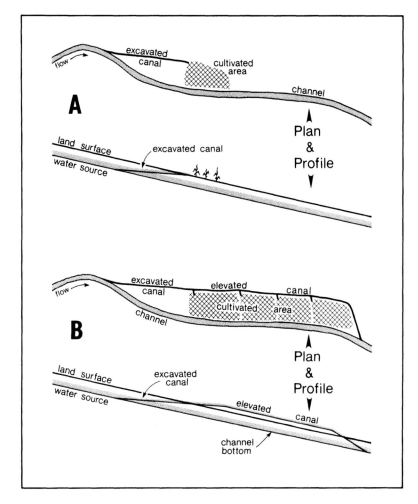

Figure 3.8. Schematic diagram of a short excavated canal system (A) and an extended elevated canal system (B).

Depths and widths ranged between 30 centimeters and 1 meter. Unlike earlier canals, these did not have sides of uniform height. Instead, the travertine was "notched" (Fig. 3.9). Although an apparently minor modification, the notching of canals was in itself an improvement in sluice gate technology. It facilitated better water control. Rather than allowing water to overflow the canal along a large segment of its length and flood the field in a haphazard manner, the notches allowed water to enter the fields at specific points.

Figure 3.9. Travertine-encrusted canal with "notched" top on the Llano de la Taza.

From there it could be diverted across the fields with much more care and precision.

Presumably water did not flow through all the canals simultaneously. The network was simply too large and the discharge of the few springs too small for such a condition to exist. Accordingly, the canals were probably back-filled with earthen materials at strategic bifurcations in order to regulate the flow of water through the system. More research certainly needs to be conducted before this point, or any others, can be substantiated.

In sum, the canal systems in the Puebla and Tehuacan valleys late in the first millennium B.C. are interesting and important in terms of their contribution to the development of irrigation technology. By virtue of being in a valley bottom and having travertine-encrusted canals, the system of the Llano de la Taza (see Woodbury and Neely 1972) has similarities to both the Amalucan canals and the earlier ones at the Hierve el Agua site in Oaxaca (see Neely 1970). This occurrence confirms, therefore, that canal technology did not always develop as the result of simple diffusion, local inventiveness, or some combination of the two. Instead, it sometimes resulted from the acceptance of different ideas, from different source areas, in concert with local developments. Apparently, the people responsible for the building and use of the fossilized Tehuacan canal system were most inventive as well as adoptive. They seem to have used the already widely accepted knowledge about small-scale canal irrigation and combined it with ideas about sluice gates learned from Oaxaca irrigators as well as with ideas about large-scale valley bottom irrigation developed further north in the Puebla Valley, and they took advantage of canals that were being elevated as a result of natural processes. The end result was the development of a most impressive system that continued to operate and be expanded for over two thousand years.

Miscellaneous Developments

There exists a limited amount of evidence of new canals for the years between A.D. 200 and 800. The paucity of data does not, however, mean that developments or advancements in irrigation technology did not occur. Indeed, five canal systems have been confirmed as having been built during this six-hundred year period, and each is important in its own right. Two of these are in the Tehuacan Valley. One is significant in that it represents an attempt to reduce one form of labor. The other is important because the canal is longer than any recorded earlier and required the development of some unusual construction techniques in order to function. Two other canal systems are near Teotihuacan; one of these involved the first confirmed relocation of a natural stream channel. The fifth system was found in the highlands of Chiapas, indicating that canal technology was beginning to be diffused over broad areas.

Cerro Gordo

Just north of Cerro Gordo in the Teotihuacan section of the Basin of Mexico, William T. Sanders found a canal irrigation network that was dated with some assurance to the Middle Horizon, between A.D. 300 and 750. It might, however, have been initiated earlier, possibly as early as 1 A.D. (Sanders 1976: 123). If it did begin this early, it must have been on a very small scale. The system certainly reached its greatest extent and maximum degree of technological complexity at the end of the first millennium A.D.

At that time canals diverted water from a large barranca between Cerros Aguatepec and Tezqueme (Fig. 3.10). One canal parallels the channel for its first 500 meters before supplying water to fields on the west of the barranca. Although it was difficult to trace any farther, it apparently continued on for another 800 meters before emp-

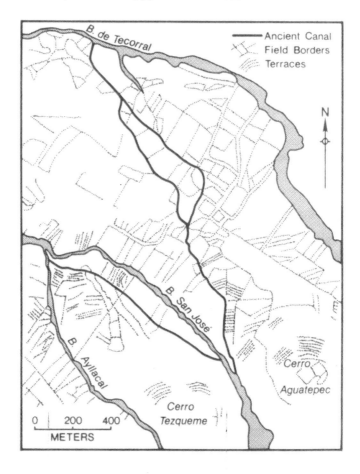

Figure 3.10. Map of irrigation canals north of Cerro Gordo. After Sanders 1976.

tying into another channel that flowed into the main barranca channel. In some places, the canal remnants appear to be between 2 and 4 meters deep (Sanders 1976: 121), but it was impossible to determine the exact dimensions because subsequent erosion and downcutting have altered the landscape greatly. Regardless of specifics, this canal appears to have been large enough to carry substantial amounts of water, even if for only short periods of time. Any traces of the diversion device have long been lost to erosion (Sanders 1976: 123). Some type of large rock diversion dam must have been used; no stake and brush device would have been able to divert the torrents that surely flowed through this steep-gradient ephemeral stream channel.

Approximately 100 meters upstream of where the west side canal intersected the main channel, another canal took water from the barranca to lands on the opposite side. As in the case of the canal on the west side, the diversion dam for this east side canal has been totally destroyed and the canal itself has been heavily damaged by downcutting. It is, nevertheless, rather distinct for its first 800 meters. The canal is less distinct through the next kilometer of its run. However, it is most evident that it split into two separate branches (Sanders 1976:121–122). The canals reconnect for approximately 400 meters before emptying into another barranca.

The fields irrigated by the east side canal are not much better preserved than the canals themselves. Indeed, not only have natural hydrologic processes modified the landscape, but historic and present-day farmers have also altered the area. From what evidence could be pieced together about this system, Sanders has concluded that the canals carried water to less than 100 hectares of fields made level by the construction of rock terraces on formerly gentle slopes. Small earthen embankments were used to divert water into the field canals.

Details on the Cerro Gordo systems are not as plentiful as those for other systems. Certainly both chronological depth and control are lacking. Nevertheless, the data from Cerro Gordo are important. From these, for example, it is seen that systems with multiple features, including terraced fields, not only can be quite complex but also can function with little organization. That these fields were "self-flooding" (Sanders, Parsons, and Santley 1979:266) indicates that the technology had been developed to such a stage that control was minimal, although maintenance was probably not.

Tlajinga

During the fall of 1980 Deborah Nichols (1982b), working on a project that involved the use of aerial photography for identifying buried features in the Teotihuacan area, found evidence of a most intriguing irrigation system. Located on a small plain in an area known as Tlajinga on the southern margin of the site is a canal remnant at least 900 meters long (Fig. 3.11). Excavations revealed that it was between 40 and 60 centimeters deep and varied in width between 1.4 and 1.6 meters at its downstream end (Nichols 1988:19). There, the top of the canal was only about 30 centimeters below the current land surface, and the canal itself was filled with fine sand and some gravel (Nichols 1982b:101). Near its head, at the Barranca de San Mateo, the canal was 60 to 70 centimeters wide and 45 cen-

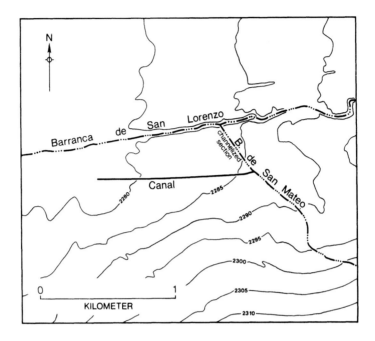

Figure 3.11. Map of irrigation canal and channelized stream at Tlajinga, near Teotihuacan. After Nichols 1988.

timeters deep. Here it was filled entirely with sand, and the top was within 40 to 50 centimeters of the surface (Nichols 1982b: 103). This canal was unlined and U-shaped in cross section.

Given the elevations of current land surface at various places, the depths of the canal, the distances from the top of the canal to the surface, and its length, it can be concluded that this canal had a gradient of 0.66 percent. This is the lowest average gradient reported thus far. In this regard alone the canal was a significant achievement in irrigation technology; however, there is more. Along its course, the gradient of the canal changed from 0.88 percent near the head to 0.52 percent near the downstream end. The decrease in gradient, whether intentional or inadvertent, had the effect of slowing the flow of water, thereby facilitating the deposition of fresh alluvium on the fields. Perhaps more important, it would have also reduced the potential damage to crops by rapidly flowing water. Coarse sediments filling the canal indicate that flooding was indeed a hazard with which irrigators had to contend. The discovery of two other canals at lower elevations in the excavations near the head of the canal and numerous other small canals that crisscrossed the stratified canals not only confirms this, but also indicates that the system had to be rebuilt repeatedly as a consequence (Nichols 1986b: 104, 103).

Nichols (1982b: 104) concluded that this canal followed the original bed of the Barranca de San Mateo, which now runs more north rather than west as it once did. She argued that the channel was rerouted to its current location (Fig. 3.11) during Teotihuacan times (Nichols 1988: 23). The presence of Tlamimilolpa Phase ceramics suggests that channelization and construction of the earlier canals might have occurred as early as A.D. 300 (Nichols 1982b: 106). The system was apparently abandoned by A.D. 750, but it was brought back into use two centuries later and used at least intermittently through Aztec times (Nichols 1982b: 106). The exact reason for the channel relocation was not addressed by Nichols. From the evidence she presents, however, it seems most likely that natural runoff across the plain was excessive and would have been detrimental to cultivation of this small area, which measured between 15 and 20 hectares (Nichols 1988: 24). Relocating the channel probably had the advantage of keeping most of the water off the plain, thereby allowing some of it to be diverted onto the fields through canals when needed. The problem, in effect, was one in which runoff agriculture, "farming in places where crops receive water as runoff from adjacent unprepared slopes" (Lawton and Wilke 1979: 4), was probably desired but infeasible. Channelizing the barranca and constructing canals from it to the fields was the next best option, and the one that was implemented.

Although the relocation or channelization of the barranca itself was for flood control and not the watering of crops, it was a significant step in the development of canal irrigation technology. This feature is the earliest existing evidence of the realignment of a natural stream channel for agricultural purposes. Earlier cultivators, for example those at Santa Clara Coatitlan, certainly channelized drainages in order to keep excess water out of their fields. What is unique about the Tlajinga channel, however, is that it marked the first time a totally new channel was excavated to do so. That the channel was excavated in concert with irrigation canals is paramount. Keeping water out of the fields in this semiarid basin made it necessary to excavate canals from the relocated channel in order to transport water to the crops. Unfortunately, the method in which water flowing into the canals was regulated remains unknown. No evidence of diversion features has been found (Nichols 1982b: 105).

Tecorral Canyon

Six kilometers south of the present-day city of Tehuacan and 2.5 kilometers west of the town of San Marcos in Tecorral Canyon, Woodbury and Neely (1972:100) found a small terraced field that was irrigated by a canal that carried water from a "reservoir" created by a "dam" across an ephemeral stream. Although it has long been breached by the stream it originally blocked, remnants of the storage dam remain visible at either side of the channel. The dam appears to have been 6 meters wide, 3.5 meters high, and 30 meters long. It was not built perpendicular to the stream, but rather diagonally across the bed, with the end feeding water into the canal being further downstream than the end on the opposite side of the channel. As a storage device, the dam was crude by standards of the day. It was constructed of an earth and rubble core, faced with dry-laid unmodified cobbles and boulders. As can best be determined, it impounded approximately 3,000 cubic meters of water.

Evidence of the canal consists of two parallel lines, 60 meters long and 75 centimeters apart, of unmodified stone slabs. No archaeological excavations were conducted, so there is no information about the canal depth. From the description of surface remains, however, it appears not to have been much different than the earliest suspected canal, that at Teopantecuanitlan. Presumably the canal would not have been much, if any, deeper than it was wide. This canal could not have carried much water, but a great deal would not have been needed. The field at the downstream end of the canal measured only about 50 by 50 meters, or approximately 0.25 hectare. Although the field appears to have been cultivated as recently as A.D. 1400, a few ceramic fragments found in association with the dam and the field indicate a late Palo Blanco Phase date (A.D. 300–700) for the principal use of these features.

Woodbury and Neely (1972:101) argue that, because the dam crossed an ephemeral stream, water might have been stored in the reservoir in order to irrigate two crops per year. This explanation seems sufficiently reasonable, but must be qualified. There is no evidence of a floodgate at the bottom of the dam such as was the case at Xoxocotlan. Also, because the bottom of the dam is slightly below the level of the fields, water drained from here could not have been used to irrigate a crop. Instead, water must have been taken out of the reservoir at the top of the dam (Woodbury and Neely 1972:100). This could only mean that water was allowed to flow into the canal when the reservoir was full. Once the level of the water in the reservoir dropped below the level of the canal, water could be taken out

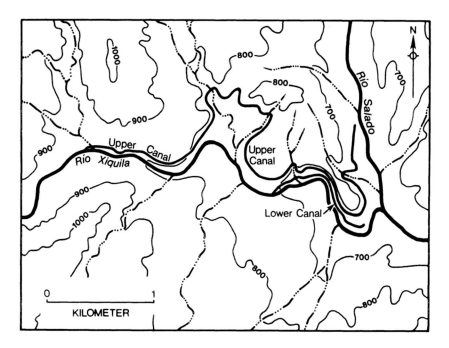

Figure 3.12. Map of Xiquila canals. After Woodbury and Neely 1972.

only by hand. In other words, the canal seems to have been built in order to facilitate a type of pot irrigation. Usually pot irrigation involves carrying water to individual plants. As a result of building a dam and canal, however, the movement of the irrigator through the field would have been replaced by the relatively simple task of standing in one place and dipping water out of a reservoir and pouring it into a canal.

Xiquila Aqueduct

Approximately 50 kilometers south of Tehuacan, at the confluence of the Ríos Xiquila and Salado, are the remains of two canals that take water from the relatively steep-graded, rapidly flowing Xiquila and irrigate approximately 20 hectares on the floodplain of the relatively low-graded, slow-moving Salado (Fig. 3.12). In at least this one respect, these canals are similar to those of the piedmont in the Cuicatlan Cañada.

The two canals actually parallel each other, one being higher up on the slope than the other. The lower one is slightly more than a kilometer long. Because of the paucity of surficial evidence, little is known of this canal. Settlements located along it, however, indicate

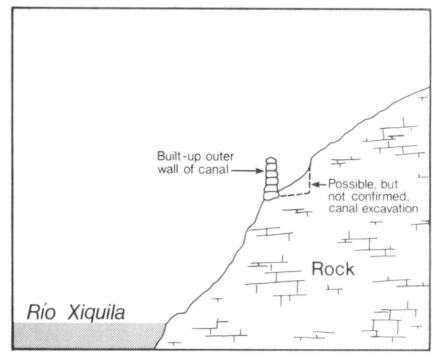

Built-up outer
wall of canal——▶

◀—Possible, but
not confirmed,
canal excavation

Rock

Río Xiquila

Figure 3.13. Schematic cross section of the main Xiquila canal.

that it was used between A.D. 400 and 700 (Woodbury and Neely 1972:109). The upper canal is later, longer, and much more impressive in terms of its material remains.

This canal was built about the same time as the lower one but was used until approximately A.D. 1540 (Woodbury and Neely 1972:109, 108). It was at least 6.2 kilometers long, and it is evident by a stone wall that formed the outside section of the canal in places where the deeply incised Xiquila left nothing but steeply sloping terrain over which the canal had to run (Fig. 3.13). According to its discoverers, who considered this canal an "aqueduct" because of its unusual walled construction along a steep slope, the masonry is of rectangular blocks that were taken from the slopes on which the canals are located. The blocks appear to be roughly dressed, but this characteristic probably is the result of natural processes; the sandy limestone is heavily folded and fractured (Woodbury and Neely 1972:105).

The actual width of the Xiquila canal is unknown because it is now obscured by materials that have fallen into the feature during post-use landslides. Furthermore, it appears to have varied in width, perhaps between 1 and 2 meters. In places where it is wide, the outer wall of the canal is quite low. Conversely, where the canal is narrow, the wall is somewhat higher. Nowhere, however, was it necessary to build the wall more than 3 meters, or ten courses, high. Given that the wall was constructed of roughly rectangular blocks, the canal was probably straight-sided or rectangular in cross section. Although no traces remain, probably because of the destructive forces of rapid runoff, aqueducts of some type must have been built in those places where the canal crossed ephemeral tributary barrancas.

Little is actually known of aqueducts in ancient Mexico prior to this time for one very simple reason: with the exception of the small ones that involved earthen fill at the earlier (300 B.C.–A.D. 200) Loma de la Coyotera site in the Cuicatlan Cañada of Oaxaca (C. S. Spencer 1982:224), none have been found. It could be argued, of course, that some existed but have not yet been discovered. While this explanation does have intrinsic merit, it is not in concordance with the available data. Of all the earlier canals for which evidence exists, none pass over large tributary streams. In every case discussed thus far, canals either begin and end entirely within the confines of tributaries (e.g., Santa Clara Coatitlan and Monte Albán Xoxocotlan) or follow the contours around the larger tributaries (e.g., the Loma de la Coyotera site). The Xiquila canal is the first that has breaks in it where it crosses tributaries. There is no evidence that masonry walls were constructed across the tributaries to support the canal. Accordingly, some type of suspended aqueduct must have been used, such that water could flow underneath it.

The construction of a suspended structure was indeed a significant achievement, however rudimentary it may have been. Although there is no way of knowing what these aqueducts looked like, they probably were nothing more than hollowed-out logs connected end to end and supported by some type of wooden trestle (Bribiesca Castrejón 1958:71). Such features were used in historic times (Fig. 3.14), are still used in places today (Fig. 3.15), and have been suggested as having been used at the Cuicatec site in the Cuicatlan Cañada (Hopkins 1984:104) ca. A.D. 1000, and elsewhere in Oaxaca (Winter 1985:104) in prehistoric times.

The diversion device used in conjunction with the Xiquila canal has long been obliterated, as have those in similar situations elsewhere. Little was actually known about ancient diversion structures at the time this canal system was investigated. Accordingly, Woodbury and Neely (1972:108) could only speculate that "A large masonry structure across the river, completely damming it, would

Figure 3.14. Hollowed tree trunks like those used for the conduit on a trellis-type aqueduct during the seventeenth century at the Hacienda Las Golondrinas (now an outdoor museum) near Santa Fé, New Mexico.

have been unnecessary, as the diversion was certainly never more than a part of the total flow, even at low water." Given its location, and what is known from recently assessed diversion features in the Cuicatlan Cañada, it now seems most probable that a diversion dam was used. Here, however, it is likely that the dam was constructed mainly out of rock rather than brush, as was the case in the Cuicatlan Cañada. The use of such features in the area today (Fig. 3.16) supports this point. Also, the presence of a rock-core dam in Tecorral Canyon nearby clearly indicates that more permanent diversion structures were being built at this time. The amount of water flowing into the canal could have been regulated by alternately backfilling and removing rubble and earth from the mouth, a technique developed at Loma de la Coyotera.

The significance of the upper Río Xiquila canal lies only in part with its impressive technology. Another important aspect of it is the energy that went into construction. While it might seem that a

Figure 3.15. Trellis-type aqueduct in use today in the Valley of Sonora. Once probably involving hollowed tree trunks, the conduit is now made of overlapping sheets of corrugated metal.

large, organized labor force would have been required, Woodbury and Neely (1972:112) calculated that it could have been built in 8,580 person-days, or the equivalent of ten people working less than three years. Combined with the fact that relatively little farmland was actually irrigated, this canal is evidence that small groups can accomplish a lot and that apparently complex features and systems need not have required a large and centrally organized labor force. Indeed, most, if not all, of the canal systems used prior to A.D. 800 were of a community nature. They were small, watered limited amounts of land, and were found in association with either small settlements or

Figure 3.16. A rock diversion dam used today in the Xiquila Valley. Stream flow is from right to left; canal mouth is in the lower left. From Woodbury and Neely 1972.

a few larger settlements. There is no direct or confirmed evidence that the canals reviewed thus far involved great amounts of labor or complex organization for either construction or use.

Lagartero, Chiapas

The last canal system dating earlier than A.D. 800 for which confirmed evidence exists was found considerably further south than any of those previously discussed. Found through an investigation of aerial photographs by Ray T. Matheny and Deanne L. Gurr (1979), traces of canals were identified in the upper reaches of the Río Grijalva drainage, in present Chiapas state near the border of Guatemala (Fig. 3.17, site 22). Ground truth verification confirmed the existence of these canals and provided insight into their function.

Water appears to have been raised and diverted out of a braided perennial stream by a dam of unreported size, shape, and construction (Matheny 1982:173). It then flowed through a main canal that was reported to be nearly 3 meters wide (Matheny and Gurr 1979:444). Whether or not this canal was as wide as claimed remains to be determined. Given that it was not excavated archaeologically (Matheny and Gurr 1983:86), the canal might well be somewhat smaller than surface evidence indicates (see, e.g., Woodbury 1960). Although the actual width and the depth of the canal are uncertain, evidence from aerial photographs clearly indicates that this feature was approximately 600 meters long (Matheny and Gurr 1979:443, Fig. 2).

Toward its downstream end, the main canal branches into a number of smaller distribution canals. Excavations of these canals indicated that they varied considerably in size, ranging in width from 80 centimeters to 1.5 meters, and in depth from 30 to 40 centimeters. Although evidence was not presented, Matheny and Gurr (1979:444) claimed that the flow of water was "controlled by gates that may have been boards." The fields themselves, however, were apparently wild flooded (Matheny 1982:173), although their sizes were not reported.

The exact dates between which this system was used are unknown. About all that has been ascertained in this regard is that many Late Classic potsherds were found in the canal, indicating presumably that it was used between A.D. 600 and 900 (Matheny and Gurr 1979:444, 442).

Little else is provided in the way of information about the system at Lagartero. Matheny and Gurr (1979:443) note only that "Major controls for perennial water sources . . . Dams, irrigation channels [sic], . . . were used extensively by the Maya in the Chiapas highlands . . ." Exactly what they base this conclusion on is currently not known, especially because no other irrigation canals are reported in the literature on that region. However, present-day crop production is hampered somewhat by a marked dry season (Matheny and Gurr 1979:442), and irrigation water carried through canal systems not unlike that identified at Lagartero contributes to agriculture in the upper Río Grijalva basin (Matheny and Gurr 1979:444). Presumably, therefore, the ancient canals would have facilitated intensive cultivation, probably multiple cropping, that would have been necessary to sustain the great number of people known to have resided in the region at the time.

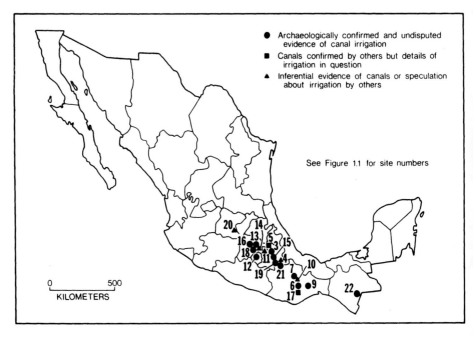

Figure 3.17. Locations of canal irrigation, A.D. 200–800.

The Lagartero canal system appears not to have involved any major technological developments over earlier known systems. It is important, nevertheless, in that it demonstrates that canal irrigation techniques developed elsewhere were becoming widespread phenomena (Fig. 3.17).

Other Possible Systems

The only other evidence for new canal systems having been built prior to A.D. 800 is sketchy at best, and involves only two areas. It is both curious and significant, however, that both cases involve broad floodplains of perennial or permanently flowing rivers. Such environments are much different from any of those where canals have been confirmed previously, and the systems might have involved large, coordinated labor forces to both build and operate.

Zaachila-Zimatlan

In the southern half of the Zaachila-Zimatlan plain of the Río Atoyac, about 15 kilometers south of Monte Albán in Oaxaca, Richard

Blanton found tentative evidence of water control works far from any tributary streams, and apparently involving the Atoyac itself (Flannery 1983 : 328). According to Blanton, the system might have consisted of canals that diverted water from the river to fields on the floodplain, where broad expanses of alluvium exist between meander loops. Unfortunately, these suspected canals have not yet been excavated archaeologically, and there is reason to doubt that they exist or, if they do, that they ever functioned adequately or were important.

Although the Zimatlan plain is considered fertile farmland, and some natives did irrigate two crops each year during the sixteenth century, water resources were apparently not sufficient for irrigation to have been practiced on anything other than a local scale (Taylor 1972 : 95–97). Complicating matters, the Río Atoyac may have been more of a liability than an asset in this part of its course. On several occasions during Spanish colonial times, and therefore presumably earlier, floods deposited thick layers of sand on lands adjoining the river, rendering them useless. Little else is known about these canals, especially details of construction or use. They are purportedly visible on aerial photographs, and sites along them were used during Monte Albán IIIa times, between A.D. 200 and 450 (Flannery 1983 : 328). More work, however, is definitely needed before their role in the development of canal irrigation technology can be determined.

El Bajío

The last suspected canal systems that might have been used during this time are in the Bajío region, much further north than any of those discussed thus far (Fig. 3.17, site 20). For the area along the Río Laja in the extreme southeastern corner of present-day Guanajuato state, where the river flowed constantly and sufficient floodplain land existed, Leonard Manrique C. (1969 : 686) suggested that "small-scale irrigation" was possible. Whether he was suggesting that canals were used prehistorically is not at all clear.

Similar confusion reigns over the possible existence of pre-Hispanic canal irrigation in the areas of what are today the cities of Queretaro and Ampaseo. The prehistory of the eastern Bajío is not understood in detail. It is, however, known that indigenous people settled in two places on the broad floodplain of the Río Queretaro: where it emerges from a canyon and enters the Bajío, near the present-day city of Queretaro; and at its confluence with the Río

Laja near what is today the town of Ampaseo. In the vicinity of Queretaro, very early during Spanish times, "A canal conducted water from the stream to irrigate small gardens and fields that produced not only the indigenous staples of maize, chile, and beans, but a variety of European fruits and vegetables introduced by the Franciscans" (Murphy 1986:89). At Ampaseo "a spring of substantial volume surfaced close to the [Río Queretaro]. . . . The site thus presented an unusual attraction [to the Spaniards]: a perennial water supply that could be easily captured for irrigation" (Murphy 1986:9). It is clear that the canals cited here were used in Spanish times. Exactly who built them, and when, however, remains unknown. The documentary materials indicate that the first canals used in the Bajío were built by Spaniards (Murphy 1986:1). That the major indigenous crops were being irrigated in the sixteenth century suggests, however, that the canals might well have been constructed, used, and, indeed, needed prior to the Spaniards' arrival. Michael E. Murphy (personal communication, 1987), a historical geographer who has recently completed a most exhaustive and thorough history of Spanish irrigation in the region, claims that it was certainly "possible" for canals to have been used in pre-Hispanic times at Queretaro and "*probable*" that they were employed at Ampaseo (emphasis mine).

The famed Mexican anthropologist Angel Palerm initiated a survey of the entire Bajío area, including the Lerma Valley and its major tributaries, the Ríos Laja and Queretaro, in the late 1970s. Herbert H. Eling continued the work after Palerm's death, completing the fieldwork in the early 1980s. According to Eling (personal communication, 1987), no evidence of prehistoric irrigation was found, even though canals were one of the items specifically sought during the course of the work. Numerous terraces, walls, and other agricultural features that can be attributed to the Spaniards were encountered, but no evidence of pre-Hispanic canals was found. There have been a number of prehistoric habitation sites dating between A.D. 300 and 900, discovered in the area, especially along the Río Laja (Braniff 1974; Brown 1985). Given the extent of historical construction, however, it might well be that the Spanish features obliterated or overlie any earlier evidence of canals. If irrigation canals existed in the Bajío prehistorically, they probably were used when the population was at its zenith, prior to A.D. 900. Furthermore, they probably would have been short, small, and not very complex. Some traces would surely remain had the canals been long, wide, and deep and carried water to more than a few tens of hectares.

The Southwest Connection

The significance of the suspected riverine floodplain canals in the Bajío lies not in their size, however, but rather in their geographical location and, therefore, the role they play in understanding the connection between ancient Mesoamerican developments and the origin of canal irrigation in the American Southwest. With the notable exception of Richard B. Woodbury (1962:303), who remained skeptical because little information was available, most scholars (e.g., Armillas 1949:91) have long held that prehistoric canal irrigation in the Southwest was probably the result of diffusion from Mexico. Emil W. Haury (1976:50) even stated most emphatically that the builders of the earliest known canal in the Southwest, the Hohokam, "were a migrated people, having moved out of the south and having been equipped with knowledge and materials theretofore unknown in the valleys of the Gila and Salt Rivers. The Hohokam seem to have known what they wanted and where they were going. An immediate and practical problem faced them on their arrival in the Southwest, and that was the digging of a canal." Haury (personal communication) still supported this argument in 1986.

There is little disagreement or doubt that Mesoamericans were long in contact with Southwestern peoples. Numerous studies, such as those by Ralph L. Beals (1932), Albert H. Schroeder (1965; 1966), Robert H. Lister (1978), and Randall H. McGuire (1980), illustrate that many Mesoamerican cultural traits were common in the Southwest. On the basis of the data now available, however, the notion that canal irrigation is one of these traits seems erroneous.

The irrigation canal of which Haury was speaking—that at Snaketown in southern Arizona—was a technological accomplishment of monumental proportions. In terms of complexity it simply had no rival anywhere in Mexico. Unlike virtually all other Mesoamerican canals that either diverted water from ephemeral streams a short distance to floodplains fields or tapped springs, the earliest Snaketown canal conveyed water over 5 kilometers from the broad, deep, and permanently flowing Gila River to fields on the upper alluvial terrace, some 15 meters above the stream channel. This canal varied in width from 3 to 5 meters, was 1 meter deep, U-shaped in cross section, and plaster-lined in places where erosion must have been a problem, at sharp turns and where water was diverted out to the fields (Haury 1976:132, 137). Although the dates of its construction are in dispute, no one doubts that it was built before A.D. 650 (Schiffer 1986:26) and no earlier than 300 B.C. (Haury 1976:149).

More recently collected evidence lends additional support to the idea that Hohokam canals were far too technologically advanced to have been the result of diffusion from Mesoamerica. Excavations conducted as part of the Tempe Section of the Outer Loop Freeway System, Maricopa County, Arizona (TSOLFS), known in the vernacular as the "Price Road Project," have uncovered the "earliest canal" in the Salt Valley (Masse and Layhe 1987:107). This canal is parabolic in cross-sectional shape, approximately 2 meters wide and 55 centimeters deep (Ackerly 1988:6.24). It had a gradient of 6.0 millimeters per meter (Masse and Layhe 1987:107) and was located not on the floodplain but on the second terrace (Masse 1987a:12). This canal was originally thought to date as early as the beginning of the Santa Cruz Phase of the Colonial Period (Masse 1987b:189), or approximately A.D. 700. Recently determined radiocarbon dates of materials taken from the canal indicate, however, that construction actually took place between 100 B.C. and A.D. 350 (T. Kathleen Henderson, personal communication, 1988).

The exact areal extent of land irrigated with water carried in the Price Road canal has not yet been determined. Calculations derived from a map drawn by Linda Nicholas and Jill Neitzel (1984:165) indicate that as many as 750 hectares may have been involved. By any standards this was a very large irrigation system. Indeed, for its time, this was the largest canal network in either North America or Mesoamerica. Furthermore, unlike earlier and contemporaneous Mexican systems that involved floodplains and sources of water above the fields, the earliest Hohokam canals watered terraces well above the water source. The engineering involved in laying out a 13-kilometer-long main canal in order to divert water out of the river far upstream of the fields was not only a monumental technological accomplishment, but one that had not been made earlier.

Given their ages, sizes, settings, degrees of complexity, and the fact that they were separated from Mesoamerican systems by over 1,800 kilometers, it appears that the Southwestern canal irrigation systems could not have had their technological origins in Mesoamerica. Diffusion through northern Mexico must be ruled out. Contrary to popular opinion, data indicate that prehistoric canal irrigation technology was developed independently in the American Southwest.

One could, of course, argue that the differences in canal technology between Mesoamerica and the Southwest were a function of either environmental adaptation (e.g., Denevan 1983) or stimulus diffusion (e.g., Kroeber 1940). The former argument seems most unlikely, as the technological developments evident in the Southwest are of too great a magnitude. Typically, adaptation involves small changes or modifications, not quantum leaps or breakthroughs. As for stimulus diffusion, it typically results in forms that are less, not more, complex than those at the point of origin.

Unlike some technologies, such as axes and hoes, that are portable and can be duplicated exactly at great distances from their point of origin, irrigation systems cannot be physically moved; only the idea can be diffused. As a result, precise knowledge of the skills involved is often lost in the process, and degenerate forms unrelated to the original types appear in places distant from the locale where the technology first arose. Given such a scenario, it seems implausible that Mesoamericans introduced canal irrigation technology that was more advanced than that which they themselves possessed!

4. A Period of Expansion and Intensification, A.D. 800–1200

On the Margins of the Southwest

The regional distribution of ancient irrigation canals in Mexico changed markedly by the end of the first millennium A.D. (Fig. 4.1). Between A.D. 800 and 1200 canals appeared for the first time in the northernmost present-day states of Chihuahua and Sonora.

Trincheras Area

In the far northwest, in the Trincheras culture area, Thomas G. Bowen (n.d.) found firm evidence of prehistoric canals near two riverine sites while conducting a surface survey supplemented with limited excavations in 1966 and 1967. On the floodplain along the south side of the Río Asunción between the towns of Caborca and Pitiquito, he identified remains of three irrigation canals that led away from the river (Bowen n.d.: 28). These canals are not sufficiently preserved to indicate exactly when or how water was diverted from the stream, but Bowen (n.d.: 56) was able to trace them on the surface for over 1 kilometer. In all probability these canals irrigated a maximum of 35 hectares each.

The canals apparently did not function simultaneously, but rather were used at different times. They were "superimposed," overlying or crossing one another, indicating sequential constructions possibly due to flooding (Bowen n.d.: 56–57). The ephemeral streams of Sonora do flood periodically, and the consequences are often disastrous. Avulsion is a severe problem there, and indeed throughout the arid northern half of Mexico. Also, sedimentation and erosion frequently alternate in destroying specific locales.

On the basis of ceramics found in association with the canals and nearby habitation sites, Bowen (n.d.: 140–141) dates these canals as

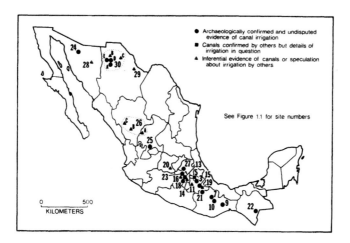

Figure 4.1. Locations of canal irrigation, A.D. 800–1200.

having been used possibly as early as A.D. 800 and perhaps as late as A.D. 1300. He also notes that their construction corresponds with the introduction of canal irrigation into other parts of the Papaguería, the extremely arid region of northwest Sonora and southwest Arizona inhabited in recent times by the Pápago Indians (Withers 1941: 49, 66–67).

Situated in a broad basin adjacent to a narrow arroyo near the location of the canals just discussed, Bowen (n.d.: 44, 57) also found remains of another canal that parallels the arroyo. He could follow it only a little more than 200 meters but noted that it was similar in appearance to the other known canals in the area. Presumably because of its similarity, he suggested that it could date as early as A.D. 800, but also suspects it was in use around the time of the arrival of the Spaniards.

Unfortunately, little else is known of the ancient canals in the Trincheras area. Indeed, that they were used for irrigation has only been inferred, and not proven (Bowen 1976: 274). From the available evidence, however, it appears that they were technologically not too well developed. Indeed, given that they were short and involved the diversion of water from low-gradient ephemeral streams, these canals are not substantially different from the canals on the Meseta Poblana, which were built 1,500 years earlier.

Evidence of canals elsewhere in the far northern expanses of Mexico prior to A.D. 1200 is sketchy as best. In some locales, such as eastern Sonora, the dates of habitation are firm but the evidence for canals is tentative. In others, such as Chihuahua, canals have been found but the dates are in question.

Figure 4.2. Stake and brush weir and a short canal that carried water from an ephemeral stream to a field near Baviácora, Sonora.

Eastern Sonora I

The sequence of agricultural development in eastern Sonora has been reconstructed on the basis of settlement and agricultural data and present-day analogs (Doolittle 1980). It was concluded that run-off agriculture involving the use of water diverted from ephemeral tributary arroyos was practiced by A.D. 1000. Numerous rock alignments perpendicular to the direction of the stream flow were found on low terraces or the floodplains of the arroyos. No firm evidence for canals associated with such features was uncovered. However, present-day farmers practicing the same type of agriculture and building similar types of rock water spreaders also excavate short canals from the arroyo channels to their fields (Fig. 4.2). It seems probable that the same type of canals would have been used pre-historically. That these canals would have been much different from those documented in the Trincheras area to the west, however, seems unlikely, as some present-day arroyo canals are rudimentary at best. Indeed, each canal carries water to only a few hectares.

Casas Grandes Area

On the other side of the Sierra Madres, in the northwestern part of present-day Chihuahua, remnants of ancient canals have long been recognized, but their ages remain unknown. Traveling in the late nineteenth century, Adolph F. Bandelier (1892:554–556) not only mentioned seeing evidence of ancient canals in the Valley of the Río Casas Grandes, but also made some elementary measurements of features visible on the ground surface. Between 3 and 5 kilometers south of the ancient city of Paquimé, he investigated remnants of a most conspicuous canal that carried water diverted from the river, which had a permanent flow (Brand 1937 : 64–67). From his account it appears that Bandelier may have been describing only the spoil banks that paralleled the canal on either side. He described the banks as having been 1 meter high on the west and 1.75 meters on the east. The canal, he claimed, was 5.5 meters wide, but this might be an exaggeration based on the distance between the peaks of the spoil banks. If parallels from other areas where canals and their spoil banks have been excavated archaeologically are applicable (e.g., Woodbury 1960), then the canal was probably as deep as the banks are high, or approximately 1.5 meters, and more than half as wide as the distance between banks, or approximately 3.0 meters. Regardless which is the most accurate, the deduced figures or Bandelier's actual measurements, the canals were exceptionally large.

Bandelier was able to trace this canal for only 800 meters, surely only a small part of its total length. At some unreported distance farther away, however, he found evidence of what might well have been a continuation of the canal. At this more distant locale, the canal measured only 4.3 meters wide or, accounting for the possibility that the distance between spoil banks was measured, approximately 2.5 meters. This narrower width suggests that some water must have been diverted out of the canal, presumably through distribution canals, before reaching this point. Bandelier also noted that this narrower section of the canal was raised slightly above the ground. Although the description is vague, it appears that Bandelier was referring to an earthen aqueduct that carried water over a natural depression or low ground (DiPeso 1983 : 17).

An assessment of currently irrigated lands, as evident on aerial photographs and topographic maps, suggests that this canal system could have irrigated upward of 1,000 hectares. Given the reported great and varied width of the canal, the height of its banks, and the perennial flow of the river through a broad, low-gradient valley, it is clear that these remains were part of a large irrigation system that was technologically more complex than any other known in the northern part of Mexico.

Much of the evidence Bandelier saw in the 1880s has been destroyed during the past century. Donald D. Brand, who began working in northwestern Chihuahua in the 1930s, conducted the first systematic archaeological survey of the region since Bandelier's.

Trained as a geographer as well as an archaeologist, Brand (1933) had a keen perception of human-modified landscapes and, knowing what Bandelier reported, he looked specifically for traces of ancient irrigation. Brand found disappointingly little. For example, in describing the canals just discussed, he noted only that canals "watered the fields of the river plain between Llano Largo and the old mission south of Casas Grandes" (Brand 1943: 142).

Elsewhere throughout the Río Casas Grandes Valley, Chihuahua, and even into parts of northeastern Sonora, southeastern Arizona, and southwestern New Mexico, Brand (1943) noted many locales near archaeological sites as being suitable for agriculture and in two cases even referred specifically to the possibility of irrigation. At the Colonia Enríquez site north of Paquimé he noted that "The soil is rich silt which could have been irrigated formerly from a less deeply grooved river by construction of brush weirs and a network of canals. No trace, however, now remains of such a presumptive irrigation system over this alluvial plain" (Brand 1943: 139). At the Loma de Montezuma site 2 miles southwest of the town of Villa Ahumada, he stated that "Excellent farm land surrounds that site and extends southwest toward the springs that *may* have been used for irrigation" (Brand 1943: 154; emphasis added).

In addition to that described originally by Bandelier, Brand could positively identify only one other prehistoric canal system in Chihuahua. That one involved a single canal that began near a spring, Ojo de Montezuma or Ojo Vereleño, at the foot of Cerro del Ojo west of the present town of Nuevo Casas Grandes, and was traced for nearly 6 kilometers. It supplied household water to the Paquimé site (Brand 1943: 141).

As was the case with the large canals on the floodplain, Brand was not the first to note the transport of water to houses via canals. In the 1550s Baltasar Obregón reported that the Spanish *conquistador* Francisco de Ibarra, saw "great and wide canals which they used to carry water from the river to their houses" (Hammond and Rey 1928: 207). That wide canals were actually used to transport water from the river, across the floodplain, to the site seems unlikely. It is more probable that Obregón confused two separate observations. Evidence suggests that some "wide" canals did carry water out of the river, but to fields, not houses. Other canals carried water to the site, but from springs and not the river.

More recent work by Charles C. DiPeso (1974; DiPeso, Renaldo, and Fenner 1974) not only has verified these earlier claims of canals within the confines of Paquimé, but also has provided additional insight. The second canal identified by Brand not only headed at a

Figure 4.3. Reservoir at the Paquimé site, Casas Grandes, Chihuahua.

spring and brought water into the heart of the ancient city for domestic use, but also emptied surplus water into canals that were used for irrigating crops in floodplain fields. The degree of engineering that went into this canal system is remarkable. DiPeso (1974: 2:344) noted, for example, that the gradient of the main canal is "delicate," dropping only 4 millimeters per meter. The canal was stone-lined and fed a series of reservoirs (Fig. 4.3) within the confines of the site (DiPeso 1974:2:344–345).

A series of smaller canals (Fig. 4.4) ran from the reservoirs to various parts of the city (DiPeso 1974:2:348–349). Near the margins of the site, these canals flowed into yet smaller canals that connected the domestic water canals to the main irrigation canal on the floodplain (DiPeso, Renaldo, and Fenner 1974:5:831). DiPeso excavated archaeologically only two of these connecting canals. One varied in width from 60 centimeters to 1.05 meters, was 2.82 meters deep, had a gradient of 22 millimeters per meter, and had stone cobble sides but an unlined bottom (DiPeso, Renaldo, and Fenner 1974:5:828). The other measured 52 centimeters wide, 23 centimeters deep, had a gradient of 19 millimeters per meter, and had both the sides and the bottom lined with cobbles (DiPeso, Renaldo, and Fenner 1974:5:829).

Neither of these canals appears to have been built with the care and degree of engineering skill that went into the building of the main canal that brought water into the reservoirs. The stone lining undoubtedly served to support the canal walls, which were under

Figure 4.4. Subfloor canal at Paquimé, Casas Grandes, Chihuahua.

considerable stress because of the construction of overlying buildings. However, the differences in morphology and the relatively
steep gradients, combined with the short lengths and the locations
of these canals, indicate that they were probably not critical components of the main irrigation system. Instead, it appears that these
connecting canals functioned principally as drains. That they emptied into canals that were used to irrigate fields was probably a
matter of convenience rather than planning.

Although the spring from which the main canal originates is
rather large, flowing at a rate of 11,400 liters per minute in recent
times and presumably in the past (DiPeso 1974:2:344), it is doubtful that irrigators with the degree of engineering expertise they apparently possessed would have built a system that required irrigation water flowing to the fields to pass through a series of reservoirs
and small canals within the confines of a large, heavily populated
site. It is more likely that such proficient engineers would have built
another canal, entirely separate but perhaps taking off from the one
supplying domestic water, for the exclusive purpose of irrigating
fields. Such a system would logically have bypassed the city. Unfortunately, no traces of such a network have been found. The reason for
this, however, might be simple enough: they have not been sought.
Although DiPeso spent years working in northwestern Chihuahua
and the Río Casas Grandes Valley specifically, he spent most of his
time working at the ancient city of Paquimé. Indeed, according to
Gloria Fenner (personal communication, 1986), who worked closely

with him, DiPeso systematically investigated and excavated no canals other than those on the site itself.

Even though he did little work on irrigation away from the Casas Grandes ruins, DiPeso did not shy away from speculating about an elaborate water control system that involved the entire Río Casas Grandes watershed. According to him (see, e.g., DiPeso 1984) runoff promoting terraces in the mountains, in conjunction with water diversion walls on the lower slopes, contributed to an increase in stream flow and, therefore, a greater and more dependable supply of water that was diverted onto floodplain fields through a complex network of canals.

The existence of such a comprehensive valley-wide system has not been substantiated. Indeed, recent research contradicts DiPeso's scheme (Schmidt and Gerald 1988). Nevertheless, numerous water control features have been found in the highlands, and canals taking water from three different types of sources have been recorded (e.g., Herold 1965; Howard and Griffiths 1966). In addition to the canals that diverted water from the perennial river and those that headed at springs, Arnold M. Withers (1963) found a canal that apparently carried irrigation water from a dam across an ephemeral stream channel. This canal was found on Mesa Escondida, west of the Casas Grandes ruins. From surface evidence it appeared to be lined with stones much like the canals found on the main site. According to Withers the canal carried water only a short (unspecified) distance to a large, open relief-free park.

Although the evidence is sparse, there is little doubt that the ancient irrigation canal systems found in northwestern Chihuahua were technologically complex. In fact, at the time, there were no canal irrigation networks anywhere in Mexico that rivaled in terms of size and technology those found on the floodplain of the Río Casas Grandes. Two major questions remain concerning these canals, however: one involves their age, and one the origin of the technology. According to DiPeso (1966:20, 21) canals began being used during the Medio Period. The earliest canals date to the Buena Fé Phase and the latest to the Paquimé Phase. DiPeso claims then that the canals were used between A.D. 1060 and 1261 (DiPeso, Renaldo, and Fenner 1974:5:828). Others, most notably and recently Stephen H. Lekson (1984), have argued that the dates DiPeso assigned to the chronological sequence are incorrect because the tree-ring dates have not been tied to the dendrochronological sequence established for the American Southwest. Those writers tend to assign later dates to the two phases in question, thereby putting the dates of canal usage between A.D. 1140 and 1400. Regardless of the dispute, however, the canals were unquestionably used between A.D. 1140 and 1261, and some were probably used much earlier, and later.

As for the origin of the technology, DiPeso was convinced it came from Mesoamerica. In one place, for example, he stated (1974:2: 336–337), ". . . Buena Fé Phase entrepreneurs arrived in the valley imbued with the age-old knowledge of hydraulic agriculture." Elsewhere, he was more specific, stating: ". . . the engineering knowledge required to produce a conservation system [that included irrigation canals] such as was introduced into the Gran Chichimeca and the Casas Grandes area after A.D. 1050, could certainly have come fullfledged from the lands south of the Tropic of Cancer where it had, for at least 1,500 years, grown into part of the cultural ethos" (DiPeso, Renaldo, and Fenner 1974:8:830).

As has already been demonstrated, Mesoamerican peoples did have various types of water control features, including canals for irrigation, prior to the development of the system used in Chihuahua. The technological complexity of the canal network near Casas Grandes, however, was much greater than any known irrigation system farther south in central Mexico. There were no other canals as large as those at Casas Grandes anywhere else in Mexico at the time. Furthermore, there were no other broad valleys with low-gradient perennial streams being cultivated on the same scale as in western Chihuahua. In contrast to earlier and contemporaneous Mesoamerican canal systems, the one at Casas Grandes was morphologically and, therefore, technologically similar to those used earlier in the Gila and Salt river valleys of Arizona. This evidence suggests, accordingly, that canal irrigation technology probably diffused southward into northern Mexico, rather than northward from Mesoamerica as DiPeso maintained.

Conchos

Ancillary data, albeit tentative and speculative, tend to support the supposition that canal irrigation in far northern Mexico originated farther to the north and diffused southward. J. Charles Kelley found abundant data indicating that Anasazi and Mogollón people, both of whom possessed knowledge of canal irrigation, began migrating into northern Chihuahua shortly after A.D. 900. They had populated large river valleys rather densely by 1400. These people settled along the Río Grande south of the present-day border city of El Paso (Kelley 1949a; 1956), near the junction of the Ríos Grande and Conchos (Kelley 1952b; 1953), and further up the Conchos, just east of present-day Chihuahua city (Kelley 1949b; 1952a).

According to Kelley, agricultural fields in these extremely arid areas are dependent on water from one of two sources: runoff from ephemeral tributaries or periodic river flooding. He noted that avulsion resulted in floodplain lands not being sufficiently stable for farming and, therefore, "made diversion irrigation impractical" (Kelley 1952a: 379). However, in a footnote appearing in the same article in which the above statement was made, Kelley also noted that there were some irrigation canals in use during the early 1950s, and that these canals were used over lengthy periods. He said: "Most of the river diversion systems of this region date back to the middle nineteenth century although some of them are as old as the Spanish colonization in the late eighteenth century" (Kelley 1952a: 358). Elsewhere, he says canals were in use by A.D. 1715 (Kelley 1986: 121). In other words, although Kelley argued that the area was unsuitable for the use of permanent irrigation, some of the canals currently in use might actually have been used continuously for more than 250 years. It is possible, therefore, that if recent and historic cultivators could use irrigation canals successfully for so long, then their prehistoric predecessors, who undisputedly possessed knowledge of the technology, could have also.

Of course, there is no confirmed evidence of this. Kelley (1952a: 366, 1952b: 263) argued that prehistoric cultivators along the Conchos did not use canals. His conclusion is based on the reports of early Spanish explorers. Those individuals made no references to irrigation in eastern Chihuahua, but they did for other areas, specifically the pueblos of present New Mexico state. For example, Antonio de Espejo and one member of his expedition, Diego Pérez de Luxán, made glowing references to the expertise of irrigators and the complexity of canal systems near the pueblos along the upper Río Grande (Bolton 1908: 178) and near Acoma and Zuñi pueblos (Bolton 1908: 183; Hammond and Rey 1929: 87, 92). They made no mention of canals, however, in their travel down the Conchos and near its confluence with the Río Grande.

Kelley's conclusion may well be wrong. That Espejo's chronicler reported canals in northern New Mexico and not along the Conchos is not sufficient evidence that irrigation was not practiced at the latter locale. Espejo and Pérez de Luxán may well have marveled at the puebloan irrigation systems because they were conspicuous, large, and provided a substantial amount of food. Along the Conchos and the lower Río Grande, in contrast, agricultural goods probably provided only part of the subsistence needs, and that proportion obtained by irrigation, if it existed, would have been very small. Furthermore, the canals themselves would have been small, and the fields would probably have covered only a few hectares and looked very much like those dependent on either runoff or river flooding. The explorers' lack of references to canals in Chihuahua might well have been due to their small size, limited economic importance, and overall inconspicuousness. Although there is no firm evidence that canals existed there prehistorically, there is no evidence they did not, and circumstantial evidence suggests they did. Certainly, however, if canals were used along the Conchos, they would have been technologically most simple and, therefore, would have played only a small, if any, role in the development of irrigation.

Throughout Terra Incognita

Until recently there was no confirmed evidence that the use of canals for the purposes of irrigation was a practice known elsewhere in northern Mexico during prehistoric times. Writing in the 1930s, Brand (1939:96) noted that "There is neither archaeological nor documentary evidence of ditch irrigation being practiced prehistorically or at the time of the conquest in Chihuahua (the case for Casas Grandes is not clear), Durango, Zacatecas, coastal Nayarit, and Sinaloa. Only at the north and south (among the Pima, Opata and prehistoric Hohokam of Sonora and southern Arizona; and on the plateau lands of Nayarit and Jalisco) do documents and archaeological investigations indicate irrigation."

The question of canals at Casas Grandes has, of course, been answered since the time of Brand's writing. Elsewhere in the region, sufficient research had been done by the 1960s to suggest that canals might have been used in some locales. Kelley (1966:104), for example, noted that canal irrigation "probably was used also in the peripheral Mesoamerican cultures, although its presence there is inferred and has not been demonstrated archaeologically."

La Quemada

In 1974 Charles D. Trombold (1976:157) confirmed Kelley's suspicions by finding, on an aerial photograph, remains of a small prehistoric canal irrigation network near the long-known and frequently studied site of La Quemada in southern Zacatecas state. He identified both a main canal that headed at an ephemeral stream and some branch canals that carried water to fields (Trombold 1977:98). Trombold does not mention having excavated the canals archaeologically. Accordingly, details of canal morphology and function are limited. On the basis of his assessment and careful inspection of the photograph (Trombold 1977:99), however, the system appears to have been technologically quite simple, consisting only of the diversion of water from the normally dry stream bed after seasonal storms, and involving only about 50 hectares.

That this canal system is truly prehistoric has not been fully substantiated. Complicating matters is the recent discovery of another small canal network that was clearly built by the Spaniards in the vicinity (Trombold 1985:247). Trombold (1985:260–261) maintains, however, that the former canals are probably prehistoric, and that they were built by migrants from Mesoamerica between A.D. 850 and 1000. If he is correct, then the limits of Mesoamerican canal irrigation are farther north than previously thought. There remains, however, an expansive archaeological "terra incognita" (Sauer 1954:554; Doyel 1979:554), especially in terms of canal irrigation, throughout northern Mexico.

The discovery of canals at a site so well studied as La Quemada raises a fundamental question. Were canals used, and, therefore, do remains of canals still exist, at other northern Mexican sites where irrigation was long thought not to have been practiced prehistorically? Of special concern are the major archaeological zones of Chalchihuites, Durango, and Zape. Agriculture was unquestionably practiced in each of these areas, which were heavily populated between A.D. 900 and 1350 (Kelley 1956:132).

Chalchihuites

J. Charles Kelley (1971:779) claimed that agriculture in the vicinity of the Alta Vista site in the Chalchihuites archaeological zone of northern Zacatecas state probably involved the utilization of runoff on the slope lands, floodplain irrigation along the valley bottoms, and, especially, spring-watered fields on the floors of side arroyos. The two former practices involved ephemeral waters; the latter, perennial flow (Kelley, personal communication, 1988). That any of these were dependent on the use of canals has not, of course, been established. Richard A. Diehl (1976:272) claimed that "small-scale irrigation" was used "to some extent." Exactly what he based this statement on, however, is unclear. It may well be he was referring to irrigation as simply the artificial application of water without necessarily involving canals. Furthermore, he may have been loosely interpreting Kelley's earlier statement. There is certainly no confirmed evidence of prehistoric canals in the area.

Although the area is known to have long been intensively farmed (Kroeber 1953:129), it is unlikely that canal irrigation was important, if it was ever employed. Two pieces of evidence support this conclusion. First, the cross-sectional shape of the valley is that of a V with only small tracts measuring a few hectares each located on the floodplain. Second, according to a local informant, canals are neither needed nor used today, and they have not been used historically because rainfall is sufficient for dry farming. If, however, canal irrigation was practiced prehistorically, it probably involved nothing more complex than small canals and diversion dams in the spring-fed arroyos, and short canals and brush weirs on the alluvial plains of the ephemeral stream. In either case, the technology was both simple and long used in other parts of Mexico. Little, if anything, was contributed by ancient farmers in this area to the development of canal irrigation technology.

Durango

It is unlikely that canals of any type were used prehistorically near Durango. This conclusion is based on two facts: there is little land suitable for such agriculture, and there is no evidence of ancient, historic, or recent canal irrigation in the area today.

The Río Tunal Valley, in which the present-day state capital is located, was certainly well populated in prehistoric times (Kelley 1956:129). However, most of these people resided in a 5-kilometer-long stretch along the river banks southwest of the modern city (J. A. Mason 1937:132). There, the valley is quite narrow, and alluvial lowlands involve fewer than 300 hectares in five discrete patches. These lands "probably were the principal fields used then as today" (Kelley 1971:788).

As for the canals themselves, there exists evidence for only one. It is on the north side near the Pueblito site (J. A. Mason 1937:132), but appears to date no earlier than the Spanish era. Furthermore, it was probably built for purposes other than irrigation. This canal heads at a diversion dam located approximately 500 meters downstream from the large retention dam built to create the Presidente Guadalupe Victoria Reservoir out of the permanently flowing river in the twentieth century. It extends for approximately 3 kilometers before disappearing beneath a large building that was probably once a mill. The canal reappears at the opposite side of this building and runs down a relatively steep slope less than 100 meters before emptying into the main river channel. There is no evidence that this canal is prehistoric or that water from it was ever used for irrigating crops.

Cultivated lands on the south side of the river are watered today by runoff from a broad mesa that is between 100 and 200 meters higher than the adjacent floodplain of the Río Tunal. Canals are not used on these lands today and presumably were not in former times. There are vast expanses of floodplain land northeast, east, and southeast of Durango city that are, however, irrigated by means of canals today. Although it is possible, given the state of technology at the time, that these lands could have been irrigated in a manner similar to that known from the Río Casas Grandes Valley, the prehistoric population of the Durango area was simply not large enough to build, maintain, and use such a complex system, must less need one. In addition to the alluvium along the river, there are numerous rainfall-dependent fields on the high lands (2,200 to 2,300 meters in elevation) within 10 kilometers to the west of the area containing most of the habitation sites. These lands might well have been cultivated prehistorically. It certainly would have been easier to dry-farm these lands than to control a perennial river in order to irrigate the floodplain.

Zape Area

Several prehistoric sites exist in the upland valleys of the Sierra Madres from Durango city north to the town of Zape, almost at the present-day border of Chihuahua state. This area lies completely within the territory occupied historically by the Tepehuan Indians (Pennington 1969). Little archaeological work has been conducted in the region recently, most having been done in the 1930s. The general settlement pattern seems to have been one in which sites were located on the tops of hills with farmland below (Kelley 1956:130). There is no reason to doubt that the ancient inhabitants subsisted principally by agriculture (Brand 1939:96). Whether they irrigated with canals, however, is not very clear.

J. Alden Mason (1937) conducted the most widespread archaeological survey of the region. Although his work reveals nothing about the prospects of irrigation, he does go to great lengths to include mention of the locations of sites in relation to water. For example, he notes springs at the sites near both Villa Antonio Amaro and Hervideros (J. A. Mason 1937:133, 136). He also mentions the permanent streams that flow past the sites at Guatimapé and Zape (J. A. Mason 1937:136, 140). The people who resided at these sites certainly used the water for domestic purposes. They also could have dug canals in order to irrigate their fields from these permanent water sources. Whether they actually did, however, is another question. Work done since Mason's first survey suggests they did not.

A	Tarahumar
B	Tepehuan
C	Yaqui Valley
D	Mayo Valley
E	Fuerte Valley
F	Southern Sinaloa
G	Parral
H	Saltillo
I	Chihuahua
J	Culiacan

0 — 500
KILOMETERS

Figure 4.5. Some locations where canal irrigation was introduced by the Spaniards.

For the area around Zape in particular, Brand (1939:79) observed chile peppers, onions, pumpkins, potatoes, and tobacco all being irrigated with both spring water and water diverted from the river. He also noted that "It is virtually impossible to determine if irrigation was practiced prehistorically" (Brand 1939:79). Indeed, the fact that some of the crops mentioned by Brand are known to have been introduced after the fifteenth century suggests that irrigation was also a Spanish introduction. Further evidence for this conclusion comes from another part of the Tepehuan region. At the Weicker site, a small prehistoric agricultural community 50 kilometers west of Durango city, Kelley and William J. Shackelford (1954:146) found remnants of a badly deteriorated "diversion irrigation system" that was unquestionably of "Spanish origin."

The mountainous area of northwestern Mexico has been sadly neglected by archaeologists. In contrast, it has long attracted researchers interested in ethnohistory. The limited work that has been done suggests that the region has a long and rich agricultural tradition (see, e.g., Bakewell 1971:59–60). Many questions remain to be answered—and, indeed, asked. One involves the use of canals for irrigation. From what is known of the area, canal irrigation appears to be a historic introduction. For example, evidence suggests that in order to grow some Old World crops, the Spaniards introduced canal irrigation technology to the Tarahumar (Pennington 1963:58), the Tepehuan (Pennington 1969:73, 245; Naylor and Polzer 1986:204, 228), and the Yaqui (Spicer 1980:30, 37; Beals 1945:5); into the

Mayo (Beals 1945 : 5) and Fuerte (Crumrine 1983 : 265, 266) valleys;
into the region of southern Sinaloa (Sauer and Brand 1932 : 74–75);
and near the cities of Parral (West 1949 : 67), Saltillo, Chihuahua, and
Culiacan (Gerhard 1982 : 22, 23, 197, 257) (Fig. 4.5). Some canals
might have been used in these areas prehistorically, of course. If they
had, they probably would have been much like their present-day
counterparts, small in size and technologically not very complex.
They also would have the greatest likelihood of having been used between A.D. 900 and 1350.

In the Heart of Mesoamerica

Tula

As has been seen, new work, sometimes even in relatively well studied areas as near the La Quemada site, can lead to new discoveries.
Such has been the case farther south in Mesoamerica proper. Working at the famed Toltec capital of Tula, in present-day Hidalgo state,
during the early 1970s, a team of archaeologists found "no direct evidence for Toltec irrigation" (Diehl 1974 : 192). Within two years,
however, another group discovered a series of canals that appeared to
have been part of an irrigated garden within a residential complex
(Peña C. and Rodriguez 1976 : 86–87). Six complete canals and fragments of five others were found paralleling each other. These canals
should more specifically be considered field canals or perhaps even
furrows because of their size, shape, and location. They are small
and V-shaped in cross section. The canals measure between 10 and
12 centimeters wide at their tops, have a maximum depth of 23 centimeters, and are approximately 90 centimeters apart (Fig. 4.6). As
best as can be determined, they were used between A.D. 950 and
1150/1200 (Diehl 1981 : 280).

The source of water for these irrigated residential gardens remains
unknown. It could be that runoff from within the site itself (e.g.,
from rooftops) was captured and diverted into the canals. As has
been demonstrated for Teopantecuanitlan and Casas Grandes, there
is certainly ample evidence that urban dwellers not only possessed
this technology early on, but had developed it to a high degree by
this time. It is also possible that these canals were either fed by
springs or were only a small part of a larger canal complex. Indeed,
both are sources of irrigation water in the valley today, with the perennial Río Tula and its tributaries being of greatest importance
(Diehl 1983 : 16).

Combining analyses of soil and present-day land use (Crespo

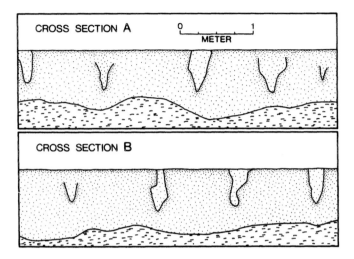

Figure 4.6. Cross section of canals at Tula. After Peña C. and Rodríguez 1976.

Oviedo 1976) with native accounts found in early Spanish documents (Feldman 1974a; 1974b) and her own investigations of traditional irrigation systems currently in use, and the locations of archaeological sites, in the Río Tula Valley, Alba Guadalupe Mastache de Escobar (1976) was able to postulate the nature of prehistoric canal networks near the Toltec capital. Although her work is speculative, it is widely recognized (e.g., Sanders, Parsons, and Santley 1976: 396), and, more important, it is most convincing.

Mastache de Escobar (1976: 64) concluded that the upstream parts of some of the canals currently in use were also used at the time of Spanish contact, and were probably used prehistorically as well. Most of these canals are located on the floodplains of the main river and its major tributaries (Fig. 4.7) and irrigate less than 200 hectares. A few, however, tap springs and divert water to even smaller plots in other locales (Mastache de Escobar 1976: 51, 65–66). Construction of the canals appears to have involved nothing more than simple excavation in the soft alluvium (Diehl 1983: 41).

Devices used to divert river water into these canals today were built by piling boulders completely across the stream channels (Mastache de Escobar 1976: 51). In effect, these features are large diversion dams. They function principally to raise the level of the water to the height of the canal mouths. There are no means by which water can be taken out during dry periods, when stream flow is greatly diminished. Although they have been long destroyed (Armillas 1949:

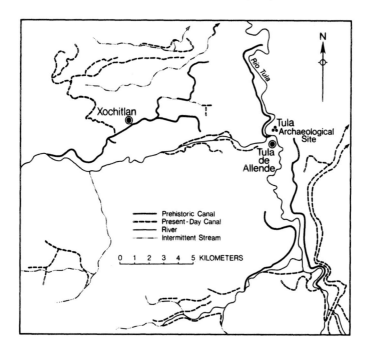

Figure 4.7. Map of Late Postclassic irrigation canals at Tula. After Mastache de Escobar 1976.

89–90), Diehl (personal communication, 1987) thinks that prehistoric diversion dams would have been somewhat smaller than the ones currently in use. The conclusion seems logical, given that some of the traditional canals being used today are longer than those probably used during Toltec times. Whatever the specifics of their construction may have been, these diversion dams show no real improvement over those discussed for the Río Xiquila near Tehuacan.

The confirmation of the use of canals at Tula does, however, contribute two things to our understanding of the development of irrigation technology in Mexico prehistorically. One is additional documentation that low-gradient, permanently flowing rivers were widely being used as sources of irrigation water. Second, and of greater importance, these canals indicate that a variety of water sources and agricultural lands (some actually within the confines of cities) were being used in densely populated areas. This last point is especially true for the Basin of Mexico.

Toward the end of the first millennium A.D., the population in the Basin of Mexico had grown so large that agricultural land was becoming scarce. Although firm evidence of changes involving canals is limited, sufficient data are available to demonstrate that new fields were beginning to be irrigated on what were unquestionably lands of marginal quality. In at least one case, this expansion involved a system that already had been in use for a long time. Another system appears to have been totally new at the time. Regardless of the specific circumstances, the bringing into use of marginal lands required a great deal of labor input as well as technological developments.

Otumba II

The system that shows clear signs of having been expanded is that near Otumba. In addition to revealing a small V-shaped canal that dated perhaps as early as 300 B.C., the road cuts found by Charlton (1977) also contain evidence of one large main canal, two smaller distribution canals, and several yet smaller field canals. The most conspicuous feature, the main canal, measured between 1.70 and 1.75 meters in depth and varied in width between 1 and 1.4 meters. For the most part it is almost rectangular or slightly trapezoidal in cross section (Charlton 1978:2, 31). The bottom, however, is not flat, as one might expect. Instead, it has the same curious stepped appearance as the canal found at Monte Albán Xoxocotlan (Fig. 2.6C).

That the unusual shape of the bottom of this canal was the result of intentional construction, as it apparently was in the Oaxacan case, seems doubtful. Charlton (1978:19, 31) reports that the entire inside of the canal is irregular and contains many small cavities; it is not smooth. It is probable, therefore, that this canal was not always maintained with the best of care and actually deteriorated over the period of its use.

In addition to this main canal showing signs of deterioration over time, there are also indications that the physical environs of the site were changing. A reconnaissance of the area with Karl W. Butzer revealed that significant changes had probably taken place since the time canals were first used at Otumba. Barranca del Muerto may not have been incised too deeply at 300 B.C. Today, however, it is downcut several meters. On the basis that a Spanish colonial roadbed runs along the floor of this channel (Fig. 4.8), it is clear that degradation occurred in pre-Hispanic times. Accordingly, in terms of geomorphology, the area today probably looks much as it did when the later canals were in use. As for its implications for canal irrigation technology, incision or downcutting of this magnitude means that the head of the main canal could not have been within a hundred

Figure 4.8. View of Otumba site looking northwest across Barranca del Muerto toward the place where Charlton found evidence of prehistoric canals. See Figure 3.2 for location where this photograph was taken. A Spanish colonial road bed lies at the bottom of the cut bank.

meters of the point where the various branches of the barranca join, but instead, must have been further upstream, perhaps as much as several hundred meters east of the present-day Otumba cemetery (Fig. 3.2).

The main canal is, of course, not visible on the surface today. Excavations revealed it to be filled with alluvial deposits. The rocks are all water-worn, but they vary considerably in size and are poorly sorted. In the lowest levels of the canal are found intermediate-sized rocks (specific sizes have not been reported). The smallest-sized rocks are restricted to the middle levels, while sediments of the largest particle sizes are in the highest levels of fill. Fine sands are found throughout the canal (Charlton 1978:19). On the basis of these sediments, it can be concluded that vast amounts of floodwater created recurring problems with the canal and, indeed, the entire system. Accordingly, it is likely that the erosional force of the water contributed to the deterioration of canal walls and bottom.

To the right of the main canal in the profile on the southeast side of the road (Fig. 3.2), Charlton (1978:31) found evidence of a distribution canal leading off from the main canal. Unfortunately he could not excavate this canal any further because it lies beneath the present-day road. It appears, however, that this canal probably carried water to smaller canals that ran across a field surface somewhere north of the present-day road.

Evidence of field canals was found during excavations on the northwest side of the road (Charlton 1979b: 22, Map 5). These specific canals did not receive water from either of the distribution canals found to date at Otumba. Although it appears from the air (Fig. 3.2) that the distribution canal evident in the profile on the southeast side of the road carried water to the field canals on the northwest side, the elevations of the canal bottoms are sufficiently different for the two to have been part of one system. The field canals evident in the northwest road cut profile could have received water only from a distribution canal further upstream and not appearing in the trenches excavated thus far. Because one is higher in elevation than the other, it can only be concluded that the canals were used at different times (Charlton 1979b: 32).

Between the field canals and the main canal evident in the profile on the northwest side of the road is a second distribution canal (Fig. 3.2). This canal clearly carried water from the main canal to fields further downstream. The fields it watered were not, however, near the end of the canal system. Indeed, they were near the head or perhaps the center of the system. Additional later canals heading from the main canal further downstream must have existed; to date they have not been discovered.

As seen in profile (Charlton 1979a: 3, Fig. 2), the bottom of the distribution canal had an elevation markedly different from that of the main canal. The deeper main canal was, therefore, capable of carrying water to several more distant fields, even at times when there was an insufficient amount to be diverted into the distribution canal. Presumably the last distribution canal in the system should have had a bottom elevation identical to that of the main canal. Near its terminus, the main canal should bifurcate, and it should be as shallow as the distribution canal evident in the road cut profile. Although the cross-sectional dimensions of the distribution canals have not been reported in the text, from drawings of excavation profiles (Charlton 1979b: 26, Fig. 5) they appear to be between 50 and 60 centimeters wide and 20 centimeters deep.

The canals evident in the road cut profiles confirm that the Otumba system not only was undergoing almost constant change, probably due to erosion and sedimentation, but also was expanding, with new lands being irrigated. The exact extent of the irrigated land probably never will be known. Given the depths and widths of the canals, as well as their degree of branching, it seems plausible that at any one time a minimum of 200 hectares could have been irrigated. Of course, the amount of land irrigated changed through time. Although change had been ongoing since ca. 300 B.C., large changes took place as the result of the rise to power of the Toltecs with their base at Tula (Charlton 1979b:33). On the basis of Mazapan Phase ceramics found embedded in the canal wall, and presumably put there during construction, the main canal at Otumba was built sometime between A.D. 900 and 1100. It continued in use, intermittently, through the Aztec Period, ca. A.D. 1600 (Charlton 1978:19).

Maravilla

Prehistoric irrigation canals have long been thought to be associated with Teotihuacan. Early in the 1950s, René F. Millon (1954), following his suspicion, conducted an assessment of the climate and demonstrated that production of a crop would be nearly impossible without some type of irrigation. Through the 1960s there was still more rhetoric about than firm evidence of canals. For example, José Luis Lorenzo (1968) argued, on the basis of known prehistoric settlements and the locations of present-day canals, that canals were used prehistorically. There is certainly nothing wrong with Lorenzo's assessment except that he provides no concrete evidence of canals.

Until the mid-1970s, when Charlton found the Otumba canals, the only confirmed prehistoric irrigation canals in the Teotihuacan area were those that were discovered, mapped, and excavated archaeologically near Cerro Maravilla just west of the ancient urban center. The Maravilla system, as it has come to be known, was a complex one that had multiple components and showed great ingenuity in construction. It differed from most other systems in that in addition to diverting water from a stream channel, it involved the relocation of the channel itself. Unlike the only other Teotihuacan channel-canal complex discussed thus far, the earlier one at Tlajinga (Nichols 1982b), the Maravilla system did not have a relocated channel that functioned to keep surplus water off the fields. Instead, the relocated channel carried water from the natural channel to a canal that then transported water to fields in what has been described as a "marginal area" (Millon 1957:160). Being in the foothills just above the valley floor, this system could not have irrigated a large amount of land, probably less than 100 hectares (Millon 1957:160). Apparently the builders of the system found it easier to divert the channel than excavate what would have been a long canal. If this was the case, then the Maravilla irrigators not only used the channelization technology that had been developed earlier in the vicinity, but also improved on it significantly. They certainly had learned a great deal about manipulating the flow of water.

An earthen storage dam 530 meters long, 11 meters high at its highest point (presumably near the center), and 7 meters wide at its widest (Armillas, Palerm, and Wolf 1956:396) was built at the upstream end of the system. This dam created a reservoir from which water could be released downstream when needed. There were no canals associated with this dam. Instead, when released, water flowed down through the natural channel across which the dam had been built (Fig. 4.9). Pedro Armillas, Palerm, and Wolf (1956) considered this feature to be of prehistoric origin. With its arched masonry and metal floodgate, however, it has characteristics of a Spanish colonial bordo (Murphy 1986:142, 149). Regardless of its age, this feature is only a minor and unnecessary component in the overall system, the most important, and clearly prehistoric, parts of which lie farther downstream.

Approximately 250 meters downstream from the dam is a smaller embankment across the channel. Although it appears to be a dam, this feature is unlike the one upstream, and it did not impound water. Measuring 45 meters in length, 2 meters in height, and 6 meters wide at the base, this rock-rubble structure redirected the stream flow in order to form a new channel along one side, rather than down the center of the small valley in which it is located. In some respects this embankment is reminiscent of the dikes used for flood control much earlier at Chalcatzingo (Grove 1984:45). There, of course, dikes were used at critical bends in order to keep the flow in the channel. The Maravilla embankment, therefore, shows a new twist to an old, but not commonly utilized, technology.

Further down this new channel, approximately 600 meters (Fig. 4.9), are the remains of a weir and a canal leading from the relocated stream channel. Armillas, Palerm, and Wolf (1956:396) claim to have found evidence of the diversion device in the form of two 25-meter-long parallel lines of round holes dug into the bedrock 1.65 meters apart, diagonally across the stream bed. The holes were reported to be spaced at an interval of between 1.20 and 1.25 meters and to have formed a zigzag pattern on the ground. Today, however, only one line of holes is visible, and these are approximately 65 centimeters apart (Fig. 4.10). Regardless of the details, brush was apparently woven in between poles that were implanted in the holes. Exactly why the people who created this system did not build a diversion dam out of stone is currently unknown. They certainly had the technology and the skill to do so had they chosen. It can only be surmised, therefore, that they wanted a brush feature. One clear ad-

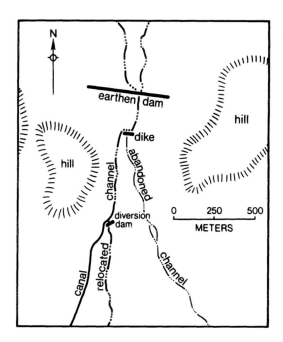

Figure 4.9. Map of Maravilla canal and channelization system. After Millon 1975.

vantage of such a device is that in times of excessive flow, such as during or after heavy storms or prolonged rainy periods, the weir would be washed away and water would flow harmlessly overland and back into the natural channel rather than through the canal and onto the fields, where damage and even destruction could occur.

The canal itself is an interesting feature. It is 750 meters long and 1.5 meters wide (Fig. 4.9). Armillas, Palerm, and Wolf (1956) did not provide details about its depth, cross-sectional shape, and gradient, but field reconnaissance indicated that it was between 30 and 50 centimeters deep and that it had either a flat bottom with rounded corners or a parabolic bottom with vertical sides (Fig. 4.10). The only other information available on this feature involves its construction. The canal was not excavated in soft alluvium as were most of those used prior to this one. Instead it was cut into bedrock, with the excavated material piled on the downslope or channel side (Armillas, Palerm, and Wolf 1956: 398). In this regard, the canal has similarities to both the Monte Albán Xoxocotlan and the Xiquila canals.

Like most other canals discussed in the vicinity of Teotihuacan

Figure 4.10. Upstream view of canal mouth and stream channel at Maravilla. Note the single line of post holes for the weir to the right of the canal.

the Maravilla system has a rather late date, A.D. 800 to 1600 (Millon 1957:164). Given what has been reported recently about channelization (Nichols 1982a; 1982b), Armillas, Palerm, and Wolf (1956: 399) were most correct when they stated that "The techniques employed in its construction may . . . be older than the system itself."

Teotihuacan Overview

The two systems discussed for the Teotihuacan area, those at Otumba and Maravilla, contribute much to our understanding of ancient canal irrigation in Mexico. They also contributed to the development of the technology. The Otumba features are perhaps the best example available to illustrate the dynamic and ever-changing nature of canals as parts of artificial fluvial systems. The stratigraphic locations of these canals, in concert with the various sizes and types of canals, also illustrate how canal systems were expanded as the need for more food to satisfy the demands of a growing population increased. These canals also indicate that ancient irrigators learned from their experiences.

In terms of technological accomplishments, the Maravilla system is the more important of the two. Its development involved the refinement of something used previously only for flood control—stream relocation.

In sum, it can be concluded that the people who lived in the Basin of Mexico toward the end of the first millennium A.D. were going to great lengths in order to bring previously uncultivated marginal land into production. Agriculture was being both expanded and intensified in order to satisfy the demands of a growing population throughout the basin. In addition to the two Teotihuacan-area systems just discussed, canal irrigation complexes dating to approximately the same time have been reported in Temascalapa and Tenayuca regions of the basin (Sanders, Parsons, and Santley 1979 : 267) and, of course, the previously discussed Cerro Gordo and Tlajinga areas, where canals were also being used at this time. Unfortunately these former systems have not been assessed in detail.

From the data we do have, one thing is perfectly clear, the people who farmed near Teotihuacan were great borrowers and adopters of technology. They discovered that in some situations, such as in the case of Maravilla, where they chose to relocate a channel rather than excavate a canal in bedrock, it is easier to manipulate nature than to replace it with something artificial.

5. The Golden Age, A.D. 1200–1520

Zeniths Everywhere

Not only were canals widespread in Mexico by pre-Hispanic times but also, to one degree or another, farmers were adopting the various technologies and expanding irrigation nearly everywhere (Figs. 1.1 and 5.1). Curiously, the time of initial development or adoption of canal irrigation and the rate at which change was taking place in any given locale were not related. In some places, people who had had no canals previously not only adopted the technology late, but built rather large and complex projects within brief periods of time. Elsewhere, farmers who had begun irrigating only a short time earlier made moderate improvements to their systems. Finally, there were irrigators who made few, if any changes, for more than a millennium. In every locale, however, canal irrigation was expanded to its maximum possible areal extent.

Regardless of the degree to which changes in irrigation were occurring in any one place, new developments in canal technology were not being made everywhere. For most areas, use of local innovations and the adoption of technology developed elsewhere, and often much earlier, were the order of the four centuries immediately preceding the arrival of the Spaniards. This is nowhere more clearly illustrated than in the case of Oaxaca.

Early farmers in both the Cuicatlan Cañada and the Valley of Oaxaca were responsible for taking canal irrigation out of its formative runoff-collection stage. They developed techniques for tapping permanent water sources, both springs and perennial highland streams. To these irrigators go accolades for "inventing" diversion dams, mortared masonry storage dams, and aqueducts, all by A.D. 200 and perhaps as early as 500 B.C. Evidence of these developments comes from Monte Albán and Loma de la Coyotera, sites that have clear dates of abandonment. The lack of direct evidence of canals after

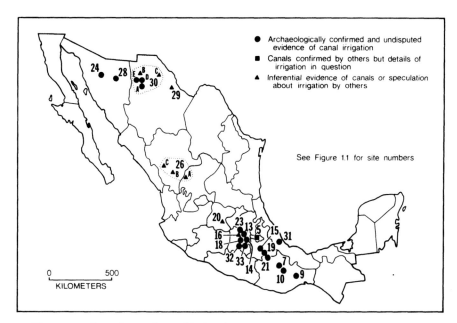

Figure 5.1. Locations of canal irrigation, A.D. 1200–1520.

this time does not, however, mean that irrigation was not practiced in the region. Indirect evidence in the form of settlement patterns (Blanton and Kowalewski 1976) and ethnographic parallels (Kirkby 1973) clearly indicates that it probably was. Although tentative, these data also suggest that new achievements were nonexistent (Winter 1985 : 106). Confirmation of this assessment comes from the Santo Domingo Tomaltepec and Cuicatlan sites.

Santo Domingo Tomaltepec

Located 1.5 kilometers north of the present-day town of Santo Domingo Tomaltepec is a prehistoric site by the same name. The site overlooks the mouth of Tsimpulatengo Canyon, through which the Río Veinte or Tomaltepec emerges. It was discovered during a survey in 1969 that this site was located in a perfect position for controlling the diversion of irrigation water (Flannery 1970). Excavations at the time, however, found no traces of canals (Lees 1970). Nearly a decade later, a detailed study of the area, including extensive excavations at the site itself, revealed only short sections of two canals. Each of the discovered canal segments measured 70 centimeters deep and 50

centimeters wide (Whalen 1981 : 212). These segments are similar in
morphology but are so far apart and oriented in such different direc-
tions that they are clearly parts of two separate canals. Although the
exact size of the area irrigated by these canals is unknown, it is unlikely, given their cross-sectional dimensions, that more than 50–60
hectares were under cultivation. Whether these canals were used
contemporaneously or at different times has yet to be determined.
Datable materials found in their association indicate only that they
were used in Postclassic times, ca. A.D. 650 to 1521.

Cuicatlan

Information on Postclassic irrigation canals is more detailed from
the Cuicatlan site than it is from Santo Domingo Tomaltepec. To
some extent, this condition might be attributable to the person
doing the research. Although Joseph W. Hopkins III is not known as
a leading figure in archaeology, those who investigate prehistoric agricultural features not only know of his work, but think highly of it.
Indeed, his master's thesis on terracing in Mexico (Hopkins 1968)
was of sufficient quality to precipitate a study of ancient terracing
throughout the New World (Donkin 1979).

The Cuicatlan site which Hopkins studied covers a long ridge
overlooking the Río Chiquito, a tributary of the Río Grande that
flows through the center of the cañada. There, Hopkins (1983:269)
found the mineralized remains of a prehistoric canal alongside a canal currently in use. Because this canal was higher in elevation than
the present-day canal and is associated with Late Postclassic ruins,
he concluded that between A.D. 1000 and 1520 residents of the valley were irrigating a larger area than farmers cultivate today, or at
least 134 hectares (Hopkins 1984:105).

The canal discovered at Cuicatlan appears on Hopkins' (1984: Fig.
8) map to be slightly more than 1 kilometer long. It carried water
that was diverted out of the Río Chiquito, probably by means of a
stake and brush weir or perhaps a diversion dam (Hopkins 1984:75).
Dimensions of the canal have not been reported but it seems to have
been approximately 1 meter deep, equally wide, U-shaped, and unlined. It had a rather steep gradient, 1.1 percent (Hopkins 1984:100)
and probably involved an aqueduct made of split and hollowed tree
trunks linked together in order to traverse a tributary stream (Hopkins 1984:104). Where space was limited, the canal was excavated
"into a hollow in the cliff above the stream bed" (Hopkins 1984:95),
not unlike what was done with the Xiquila canal.

As in the case of the Xoxocotlan canal at Monte Albán, the one at the Cuicatlan site carried water across a ridge. There, its downstream side was open in several places in order to let water flow down the slope and irrigate a series of at least twenty-six terraces, each of which was more than 3 meters high (Hopkins 1984:114). If present-day practices are anything like ancient ones, the openings in the canal were closed with rocks when water was no longer needed on a particular terrace (Hopkins 1984:77).

There were probably canals not unlike this one throughout the cañada in Late Postclassic times. In addition, there must have been canals carrying water from the Río Grande to floodplain fields (Flannery 1983:337). Such canals, however, have not yet been found. Regardless of the number and kinds of canal systems operating in the cañada, Late Postclassic evidence indicates that, as in the case of the Santo Domingo Tomaltepec site, no substantial improvements had been made in canal technology anywhere in Oaxaca since Formative times, ca. A.D. 200.

Although the dates of initial adoption are different from those at which canals were first used in Oaxaca, essentially the same conclusions about developments in irrigation technology can be drawn about places as far away as eastern Sonora and coastal Veracruz.

Eastern Sonora II

In eastern Sonora, rudimentary canals were used to carry water from low-gradient ephemeral stream channels to small plots by A.D. 1000 (Doolittle 1980). It is not clear when the technology to tap perennial streams started being used. By the time the Spaniards arrived, however, floodplains of both small, permanently flowing streams such as the Río Chico (Pennington 1980:355), a tributary of the Río Yaqui, and larger streams such as the Ríos Sonora, Sahuaripa, and Moctezuma (Doolittle 1984b:252; 1988) were irrigated to their maximum possible extent.

Details of ancient canals are, of course, lacking. Such features were, however, probably no different than those used in the region today. Present-day canals are of two different types, main canals that carry water from the rivers and distribution canals that transport water from the main canals to the fields, encompassing approximately 150 hectares. Main canals are on the average between 1.5 and 2 meters wide and 1 meter deep. They take off from the river by means of either stake and brush or earthen diversion dams (Figs. 5.2 and 5.3). Both types of structures are temporary and have to be

Figure 5.2. Stake and brush diversion dam on the Río Sonora, near Sinoquipe, Sonora.

Figure 5.3. Low earthen diversion dam on the Río Sonora, near Arizpe, Sonora. Reprinted with permission from *Geographical Review* 70 (1980): 336.

rebuilt regularly, usually annually at the onset of the dry season. These features facilitate irrigation only during times of greatly diminished stream flow. During most of the year, and certainly during the rainy summer months, water enters the main canal without any type of diversion device.

Distribution canals are smaller than the main canals. They typically measure approximately 1 meter wide and 50 centimeters deep. Unlike the main canals, these do not contain water year round. Instead, each carries water only when the irrigators shovel out the earth used to close off the canal. Once water is allowed to enter a distribution canal, it flows directly onto a field, which is free-flooded. Each field is enclosed by a low earthen bund designed to retain water. Low rock terraces and water spreaders made of brush are commonly used to distribute water evenly across fields.

Although details of the canal networks used to irrigate the floodplains of eastern Sonora rivers in pre-Hispanic times remain unverified, the technology used today has a long antiquity in Mexico. Irrigators in the region certainly improved their irrigation systems by adopting technologies developed elsewhere as they expanded cultivation into previously unused areas (Doolittle 1980). They did not, however, make any new developments or contribute to the improvement of canal technology.

Zempoala

Unlike the case of eastern Sonora, where archaeologically verifiable remains of irrigation canals have yet to be identified, there is some concrete evidence of canals at the site of Zempoala near the coast of the Gulf of Mexico in central Veracruz state. There, in a semiarid locale along an otherwise tropical coast (Lauer 1973), was found "the only irrigation system known from the Totonac area" (Palerm 1955:33; see also Sanders 1971:548). Although the Spaniards are purported to have observed irrigation being practiced at Zempoala in the sixteenth century (Kelly and Palerm 1952:99) and archaeologists have long known of the canals, there have been no systematic and detailed studies of irrigation at the site. Some writers have claimed that there existed an "elaborate irrigation system at Zempoala" (Sanders 1971: 547; see also Wilkerson 1983:58). Such conclusions, however, are based entirely on the findings of one small study that did little more than uncover an urban drainage system that "emptied into house-cisterns and then discharged into irrigation canals" (García Payón 1971:538) and an early report saying only that prehistoric canals "have been widened and are in use today" (Kelly and Palerm 1952:99n.28).

Although definitive, these comments are all too brief. They may tend to inflate the importance of the system, and, as has been the case with water control features near El Tajín, the so-called irrigation "canals" may actually be channels associated with raised fields. Impressive elevated canals do carry water to fields in the area today. However, according to a local informant, maize and beans are easily grown during the summer months without irrigation, as rainfall then is more than adequate. The canals are used principally to irrigate sugar cane during the dry season, which extends from February through early May. Given that sugar cane is a historic introduction, it may well be that irrigation canals are as well. Lending additional support for this possibility are the locations of the present-day fields themselves. Sugar cane is grown not only all around the site but also *within* its confines. There is no evidence of crops having been grown at the bases of pyramids anywhere in Mexico during pre-Spanish times. Accordingly, it is unlikely that such was the case at Zempoala.

In sum, it can be surmised that ancient irrigation canals, if they existed, might not have been all that "elaborate" and would have watered only a few hectares. The drains are certainly not very impressive when seen in the context of others. Photographs of these features (García Payón 1971:538, Fig. 32) show a great deal of similarity to the canal at Teopantecuanitlan, the earliest suspected canal irrigation site in Mexico, and the drains at the Olmec site of San Lorenzo. They appear to be slightly less than 1 meter wide and 1 meter deep. Thick, uncut slabs of stone were used to line the bottoms and sides and cap the tops. According to reports, the entire system was built in the fifteenth century A.D. (García Payón 1971:538) and abandoned after the arrival of the Spaniards (Sanders 1971:547).

As rudimentary as the drains were and the other features might well have been, the Zempoala system is nearly insignificant in understanding the development of canal irrigation technology. Clearly, the building of these features contributed little if anything to the advancement of the technology. Even if it was as large and as elaborate as some claim, this system was one that involved only the application of technology developed elsewhere and much earlier. What is important, however, is how rapidly the technology was applied once it was needed. Having been built and used within a century suggests that this agricultural water control network was not developed locally. Instead, the builders of Zempoala system were learning a great deal from people with whom they maintained close ties and

who were making great strides in the advancement of canal irrigation technology. Those people resided in the Basin of Mexico.

Monumental Works in the Basin of Mexico

By Late Horizon times, ca. A.D. 1350, the demand for food was so
great throughout the Basin of Mexico that marginal lands were not
only farmed but cultivated intensively. Accordingly, every major
source of water in the basin was being used for the irrigation of crops
(Sanders, Parsons, and Santley 1979:114).

Most of the canal networks appear to have been small, involving
less than a few hundred hectares, not interconnected or dependent
on one another, and serving only the lands of individual settlements
(Sanders, Parsons, and Santley 1979:263; Fuentes Aguilar 1988:10,
12). Water was typically conducted directly from springs or small
ephemeral streams to fields. Storage devices such as dams were relatively few and, even when large, were not too elaborate. The technology tended to be simple, involving the use of diversion dams made of
earth, rock, and brush. Canals were for the most part short, shallow,
narrow, and unlined. Accordingly, Sanders, Parsons, and Santley
(1979:252) were probably close to being correct when they assessed
these systems as a "distinctively inefficient use of a resource."

Although most systems were small and of apparently poor quality,
a few reasonably large irrigation works were used in every part of the
basin except for the Ixtapalapa area, where water sources were inad-
equate (Blanton 1972:1318, 1322), Palerm (1973:20) once claimed
that some canals were 15 to 20 kilometers long and lined with stucco
or stone. The exact canals to which he was referring are not reported. Such systems, however, must have involved considerable
amounts of labor (Sanders 1976:116). They also involved significant
technological developments. Without doubt, the greatest accomplishments in hydraulic agriculture in prehistoric Mexico were the
channelization or redirecting of rivers and the building of aqueducts
during this time.

Cuautitlan

Most of the archaeological evidence of canal irrigation that remains
today in the Basin of Mexico is, like the Maravilla system, found on
the slopes or foothills of the surrounding mountains. Little material
evidence remains of canal systems used prehistorically to irrigate
the broad plains of the basin floor because those areas were occupied

by the Spaniards, whose later activities obliterated many features, and also because of destruction due to the expansion of present-day Mexico City. There is, nevertheless, documentary data that provides some insight into valley-bottom systems used at the time of Spanish contact.

A total of twenty-two such Spanish documents, ranging from the letters of Cortés to early histories, have been gleaned by Palerm (1973) for evidence of water control, including canal irrigation. The documents perused in his study referred to nearly one hundred locales in which some form of hydraulic works was used. For the most part, only brief statements mentioning the existence of features such as canals, dikes, dams, and aqueducts were found. A notable exception was *Los Anales de Cuauhtitlán* (1945), included as part of the *Chimalpopoca Codex.* That document is important in that it contains much information about the channelization of one very large river for the purposes of irrigation. Written in the middle of the sixteenth century A.D., it discusses events that occurred in the far northern portion of the Basin of Mexico during the preceding century (Rojas Rabiela 1974:85).

From this document, it appears that farmers who were cultivating perhaps as much as one thousand hectares along the westernmost edge of Lake Zumpango, an area also known as Lake Citlaltepec, during the early fifteenth century were struck with water shortages (Palerm 1973:238–239). Whether these farmers were diverting water onto their fields from streams that otherwise would have flowed into the lake, were cultivating raised fields within the lake proper, or were doing both is not at all clear. It really makes no difference, however, as the result of diminished rainfall and runoff would have been the same.

Lake Citlaltepec is fed by only one perennial or constantly flowing stream, the Río Tepotzotlan. All others that drain the surrounding slopes flow only seasonally. Given the limited amount and periodicity of surface water, even small decreases in the amount and timing of rainfall would have significantly reduced the water available for irrigating fields adjacent to the lake. Farmers cultivating raised fields within the lake would similarly have felt the effects of drought. Although raised fields typically do not suffer greatly during dry periods, those that might have been used in the far northern part of the Basin of Mexico would have been located in the least ideal of locations. Being the uppermost and shallowest of the interconnected bodies of water, Lake Citlaltepec would have been the one most affected by a prolonged dry spell. It would have been the first to experience any drop, and the last to benefit from any rise, in the water level (Strauss K. 1974:144–146). Perhaps not surprisingly, therefore, and regardless of their cultivation technique, farmers along the margins of the lake sought a solution to the potentially devastating problem of a variable water supply.

Increasing the amount of water in any region is no simple matter. Traditionally there has been one way of doing so that seems to have been preferred over all others—divert water from one river into another (see, e.g., Ortloff, Moseley, and Feldman 1982). This option was the one chosen by farmers in the far northern part of the basin (Fig. 5.4). There, according to *Los Anales de Cuauhtitlán*, the Río Cuautitlan, which originates near the head of the Río Tepotzotlan but flows eastward to Lake Xaltocan, was diverted northward, west of the then site and now town of Cuautitlan, through a newly excavated channel over 6 kilometers long before emptying into the Río Tepotzotlan, which itself was deepened, widened, and straightened for more than 10 kilometers (Strauss K. 1974:Esquema B, p. 171). Although a major modification of the region's hydrology, this one appears not to have had the negative impact that is often associated with such large-scale environmental alterations. Indeed, it had all the anticipated benefits and more.

Prior to channelization, the Río Cuautitlan emptied into Lake Xaltocan, downstream in the basin system from Lake Citlaltepec. This river certainly contributed to the overall level of water in the basin, but it had only a minimal and indirect effect on the level of Lake Zumpango. Diverting the river northward into Lake Citlaltepec did not affect the total amount of water in the lacustrine system, but it increased the level of the water in the uppermost section. Increased water would have benefited raised-field cultivators and irrigators alike.

In addition to ameliorating the negative affects of prolonged dry periods on agriculture in the far northern part of the basin, the channelization of the Río Cuautitlan also protected the ancient town of Cuauhtitlan against flooding (Sanders, Parsons, and Santley 1979: 270) and resulted in the creation of approximately 1,000 hectares of irrigated land in the vicinity. According to the principal documentary source, additional canals were excavated at approximately the same time that the stream relocation channel was dug (Strauss K. 1974:147–154). These included one rather large canal that carried water from the relocated channel (Sanders et al. 1982:8) to a series of at least four and perhaps five other smaller canals that irrigated newly established fields near the town (Fig. 5.4). Exactly how this system functioned is not clearly understood at this time. Indeed, it has long been recognized that archaeological investigations with the

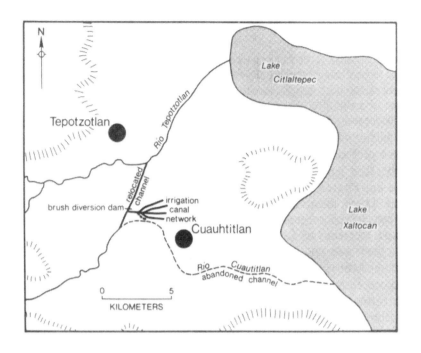

Figure 5.4. Río Cuautitlan channelization and irrigation system: A, map (after Strauss K. 1974; Sanders 1981); B, view of the channelized river as it appears today. Near the point where it is crossed by Highway 57, the channel shows signs of having been maintained since Aztec times.

intent of assessing this irrigation system are needed (Rojas Rabiela 1974:86n.139).

Details of the entire channelization and irrigation works remain scarce. As is often the case with chroniclers' accounts, *Los Anales de Cuauhtitlán* does not contain a great deal of the type of information needed here. It contains nothing about the width, depth, gradient, cross-sectional shape, and other details concerning the technology involved in either the new channel or the irrigation canals. About all that is reported is information on the structure built to direct the Río Cuautitlan into its new channel and the one that diverted water into the irrigation network.

Excavation of the new channel began sometime around A.D. 1435 (Rojas Rabiela 1974:85) and took seven years to complete (Palerm 1955:40). Sometime toward the end of this period construction was begun on the device that closed off the old channel and redirected the flow of water into its new course. This feature "was made of beams, joined and upright" (Palerm 1955:40). It must have been of herculean proportions, as it had to withstand the force and erosive pressures of a perennial stream, and it took a large, organized labor force two years to build it (Rojas Rabiela 1974:86).

The device used to divert water out of the relocated channel and into the canals that irrigated fields near Cuauhtitlan was nowhere near as impressive as the structure built to reorient the river channel. In fact, it was a comparatively flimsy structure that had to be rebuilt every year, presumably at the beginning of the dry season, when the river flow was greatly reduced (Strauss K. 1974:147–148). Although made of brush and mud, this temporary structure was not a weir but a diversion dam. Its function was "to draw and raise" water into the canals (Strauss K. 1974:149).

It is curious that an irrigation system as elaborate as that at Cuautitlan would have included such a rudimentary diversion structure at its head. According to Rafael A. Strauss K. (1974:153), however, nothing greater was needed and, indeed, anything more substantial would have been detrimental. Although the river flow was greatly diminished during the dry season when irrigation water was needed, there was apparently a sufficient volume to fulfill the needs of the farmers. Anything greater than a temporary diversion dam would have required much more work than was necessary. Furthermore, construction of a larger and less permeable device would have impounded water that was needed to maintain either irrigation near the mouth of the Río Tepotzotlan or the level of the water in Lake Citlaltepec.

In sum, the building of a channel to redirect the flow of Río Cuautitlan not only facilitated irrigation of some unknown variety along the northwesternmost margins of the ancient lake system in the Basin of Mexico but also led to the opening of a large-scale canal-irrigated field complex near the Cuauhtitlan site. In addition to being a dual-purpose project, this construction effort was both an elaboration of technology developed earlier in the basin (Sanders 1981:192) and one of the more spectacular achievements of prehistoric Mexican irrigators. The system clearly was an upgrading of the technology employed in the Maravilla canals. Whereas that irrigation complex involved the redirection of a natural channel only a few hundred meters, the Cuautitlan project involved several kilometers, two separate drainage areas, and an entire lacustrine system.

The Cuautitlan channelization and irrigation effort is important not only for its engineering properties, but also in terms of its planning. Earlier systems were, as far as the extant data indicate, small in scale and scope during the initial years of their existence. They increased in both size and complexity through either systematic or incremental change (e.g., Doolittle 1984a) over several decades or centuries of use. None of the previously discussed canal complexes show convincing signs of having been planned in their entirety. Even if they had been, however, the planning would have been on a comparatively small scale, with the laying out and construction being undertaken by the irrigators themselves.

In contrast, the Cuautitlan project must have involved people who can only be considered as true professional civil engineers. These were not conservative, provincial cultivators who accepted new ideas cautiously and, therefore, slowly. They were also not progressive farmers whose inventiveness was based mainly on experience gained in the course of cultivating their own fields. Those responsible for the planning and construction of the Cuautitlan channel and canal irrigation systems were cosmopolitan scientists who had at their disposal a great deal of accumulated information (Palerm 1955:39). The depth of their understanding was perhaps eclipsed only by the breadth of their knowledge. The effort of this system's developers was neither one of accidental discovery nor trial and error. Instead, it involved an in-depth comprehension of the physical environs of the Basin of Mexico, hydrologic principles, agronomy, and engineering.

Chapultepec

In addition to possessing the ability to manipulate successfully the hydrology of their environment, some people who resided in the

Basin of Mexico during pre-Hispanic times were also master agricultural architects. Not only could they relocate river channels *on* the ground, but they could also construct features that allowed for water to be diverted *above* the ground. This ability is nowhere more clear than in the case of the aqueducts built in order to transport water over various environmental obstacles.

The aqueducts that were built in the Basin of Mexico during pre-Hispanic times were so numerous and outstanding that they were one of the first indigenous engineering accomplishments that the Spaniards noted. Indeed, in his second letter from Mexico, and his first after being in the basin, Cortés commented repeatedly about their existence, functions, and importance (Morris 1928). Of all the aqueducts that have been reported, however, one in particular has received significant attention by scholarly researchers. The aqueduct ran from Chapultepec Hill (in what is today the large urban park that bears the same name) to Tenochtitlan, the capital city of the Aztecs, on an island in the western part of Lake Texcoco.

This aqueduct carried water that, according to legend, gushed out from under the base of a large rock (Brundage 1979:48). The flow was apparently so great that the spring long attracted people, including the Toltecs, who made this site their home after Tula was destroyed and abandoned in the middle of the twelfth century A.D. (R. E. W. Adams 1977:229, 241). It also resulted in the place being sanctified, with the Aztecs posting a priest of Tlaloc there (Brundage 1979:140).

Detailed investigations of the site have neither confirmed nor overturned the legend. They have, however, provided some additional insight. For one thing, there was not one spring but a number of springs connected by a series of canals (Braniff de Torres and Cervantes 1967:265–266). The canals in turn, carried water to some royal baths that were carved into solid rock near the base of the hill (Fig. 5.5; Braniff de Torres and Cervantes 1966:161, 164, Lámina 1). Because this locale attracted people for centuries, perhaps even millennia, there is no way to determine exactly when the springs were first enlarged. Ceramic data indicate only that it was sometime during Aztec times, A.D. 1150–1519 (Braniff de Torres and Cervantes 1966:168). The matter might be more complicated, however, as the springs, canals, and baths could have been enlarged on several different occasions. There is clear evidence that the aqueduct itself was rebuilt at least once.

According to the Mexican engineer José Luis Bribiesca Castrejón (1958:78), who investigated this feature in detail, construction began on the first aqueduct in A.D. 1418. Information has not been pro-

Figure 5.5. Spring-fed baths built by the Aztecs at the head of the aqueduct that carried water from Chapultepec Hill to Tenochtitlan.

vided on the nature of the aqueduct as it crossed the land area before heading out over the lake. Details on the part that traversed the water are, however, abundant. Construction of this section involved the weaving of reed mats of unreported lengths but measuring between 7 and 8 meters wide. These rafts were then floated into place and, once in their correct locations, anchored with stakes driven deep into the lake bed. With their lateral movements thus minimized, mud, rocks, and sod were loaded onto the rafts. Materials were continually added until the rafts sank to the bottom and the newly created lands were well above the water level. The earth-covered mats were not laid end to end in one continouus strip as one might expect. Instead they were spaced 3 to 4 meters apart, thereby creating, in effect, a chain of islands.

A conduit to carry water was built on top of each island by first piling additional mud into a linear mound approximately 1 meter high, 1.5 to 2.0 meters wide, and running the entire length of the island. This mound was reinforced by stakes driven through it and into the recently completed base. Its top was then hollowed out, and the resulting trough was lined with compacted clay to reduce seepage (Fig. 5.6A).

In order to connect the troughs between islands, the builders used

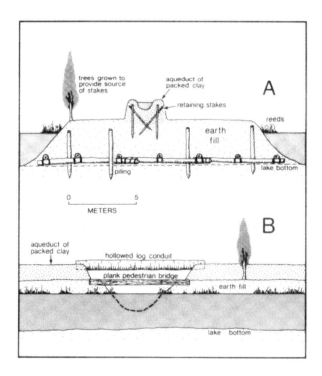

Figure 5.6. First aqueduct built from Chapultepec to Tenochtitlan. After Bribiesca Castrejón 1958.

split tree trunks that had been hollowed out. They also connected the islands with wooden planks laid next to the aqueduct. These planks allowed pedestrians to cross the lake on the same structure used to transport water (Fig. 5.6B).

This aqueduct functioned until A.D. 1449, when it was destroyed by a flood. It is not known whether it was repaired or abandoned, but in A.D. 1465 a completely new aqueduct was built to replace it. The new aqueduct may have used remains of the original islands built nearly a half-century earlier. If so, they were enlarged considerably. The islands for the new aqueduct were between 10 and 12 meters wide and tens of meters long. They were also higher above the lake level than the earlier islands (Fig. 5.7).

Although these newly renovated islands were longer, wider, and higher than their previous counterparts, the most impressive part of the new aqueduct was the superstructure that actually carried water. It was a masonry structure 1.6 meters high and 2.5 to 3.0 meters wide across the top. It was somewhat wider across the bottom, as it

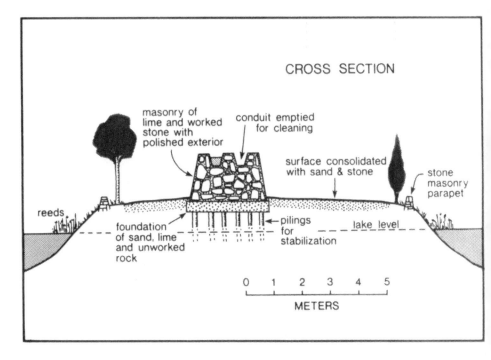

Figure 5.7. Rebuilt aqueduct between Chapultepec and Tenochtitlan. After Bribiesca Castrejón 1958.

was trapezoidal in cross-sectional shape. Rocks used in the aqueduct do not appear to have been either cut or faced. They were, however, held together by a lime-based mortar that not only covered the small stone fragments used to fill the chinks but was also used to cover the entire surface. The superstructure stood on a foundation of sand mixed with lime and small stones, and footed with stakes driven into the islands. Hollowed logs were still used between the islands.

The most curious aspect of the new aqueduct is undoubtedly the part that actually carried water. Unlike the earlier clay-lined conduit, the new one was mortared like the rest of the structure. Also, instead of one conduit, there were two, each being 75 centimeters deep, 30 centimeters wide at the bottom, and 60 centimeters wide at the top (Fig. 5.7). According to various accounts, not the least of which was by Cortés himself (Morris 1928:92–93), water was carried in one while the other was being cleaned and repaired.

Exactly what was done with the water once it reached the end of the aqueduct is not at all clear. Bribiesca Castrejón (1958) is one of many scholars who assume that it was distributed through a series of small canals and used for domestic purposes. Palerm (1955:38), however, questions this interpretation on the basis that more water than was needed for household use was brought over on it and at least one other aqueduct (Fig. 5.8). He suggests, therefore, that while some of it undoubtedly was used for drinking and washing, the bulk was used for agriculture. Most of the water was probably used to maintain the lake level around the 500 to 1,000 hectares of raised fields that existed on the edge of the island city (SARH 1979:17–21). Some, however, was used for irrigating gardens within the city (Palerm 1955:38).

The development of the earlier aqueduct was a significant technological achievement. The use of aqueducts has an unclear history

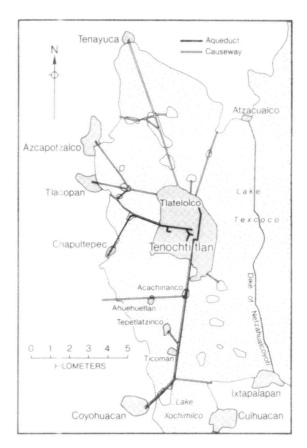

Figure 5.8. Map of western section of Lake Texcoco, Tenochtitlan, and aqueducts. After Sanders 1981.

in Mexico, but it is known that small ones were used almost 1,500 years earlier than the first one built at Chapultepec. Although there were additional developments during the interim, the first Chapultepec aqueduct was considerably longer and more complex than anything seen previously. Furthermore, it was the combination of technologies developed in four different places at four different times for four different purposes. The building of islands in the manner described might have its roots in the building of raised fields (see Wilken 1985). Conduits built to carry water across the tops of the islands were not unlike earlier canals (e.g., those in the Tehuacan Valley) that increased in height as the result of the accumulation of travertine. Bridging gaps by means of split and hollowed tree trunks originated, as best as the data reveal, 1,000 years earlier (e.g., with the Xiquila canal). Finally, lining the canals with compacted clay appears to have been developed here during attempts to repair areas eroded by the extremely rapidly flowing water that was carried in the aqueduct.

As remarkable a technological development as this aqueduct was, its rebuilding in the middle of the fifteenth century was equally phenomenal. Obviously, the builders had learned something from the flood. The rebuilt aqueduct was much more massive than the earlier one. Not only was it larger in every respect, but it was constructed with permanence in mind. Rocks replaced earth as the principal material, and mortar replaced wooden stakes as the binding agent. The latter also replaced compacted clay as conduit lining.

The use of masonry was without question a technique adopted from the building of the pyramids and temples that dominated the city. Although hollowed logs continued to be used to bridge the spaces between the islands, these were fewer in number than before, and they could certainly have been replaced in a short time with only minimal disruption of the flow.

It has not been reported, but the residents of Tenochtitlan must have had problems keeping the original aqueduct clean and maintained. A blocked conduit the size of this one would surely have taken a long time to clean and repair. Disrupting the flow for such purposes would have been most inconvenient. The building of two conduits during reconstruction alleviated maintenance-related disruptions.

Overall, therefore, the Chapultepec aqueducts were major developments in canal irrigation technology. Whereas the earlier one reflects a great many borrowed ideas, the latter one shows local refinements and improvements. That the rebuilt aqueduct was somewhat taller than its predecessor should not be overlooked as being insignificant or considered only a minor improvement. Although not immediately obvious, the ability to carry water at high elevations above the ground allowed irrigators to transport water greater distances and over extremely rugged terrain. This is nowhere more evident than with developments that were taking place on the east side of the basin in the Texcoco area.

Cerros Tetzcotzingo and Purificación

Like most other places in the Basin of Mexico, the Texcoco area was undergoing some dramatic environmental modifications during Aztec times (Wolf 1959:132; Palerm and Wolf 1961:283, 285). Numerous springs such as the one located at the base of a hill just north of the present-day town of Tepetlaoztoc (Barbara J. Williams, personal communication, 1988) were tapped for the first time during the fifteenth century. Water from these sources was collected into main canals and carried to the plains on the edge of Lake Texcoco (Sanders 1982:7), where several hundred and perhaps thousands of hectares were cultivated. There it was distributed to the fields through intricate networks of canals (Williams 1984). In many cases, springs existed high on the sides of steeply sloping hills where agriculture was infeasible, while neighboring hills with more gentle slopes of high agricultural potential had no water. The solution to this problem involved the construction of aqueducts between hills.

Documentary sources indicate that numerous aqueducts once existed in the Texcoco area (Palerm 1973). Many of these aqueducts, such as the one described in the documents as being near the town of Tezoyuca (McAfee and Barlow 1946:114–115), have long been destroyed (Wolf and Palerm 1955:271). Others, such as one measuring 400 meters long, 7 meters wide, 2 meters high, and located in the federal park El Contador near the present-day town of Texcoco (Parsons 1971:94–95), still remain but have not been studied in detail. Constructed of earth and rock rubble in Aztec times, this particular aqueduct once carried water brought through canals from as far away as Teotihuacan (Palerm and Wolf 1961:286) to fields of exotic vegetation in the royal gardens at Acatetelco (Pomar 1941:54). Appropriately enough, the aqueduct is very close to the International Center for the Improvement of Corn and Wheat (Centro Internacional del Mejoramiento del Maiz y Trigo, CIMMYT), where research into new varieties of crops is currently underway.

Undoubtedly the most notable aqueducts, and the ones that provide a wealth of information about advancements in canal irrigation technology, are the four associated with the systems involving cultivation on Cerro Tetzcotzingo and Cerro Purificación (Fig. 5.9). Lo-

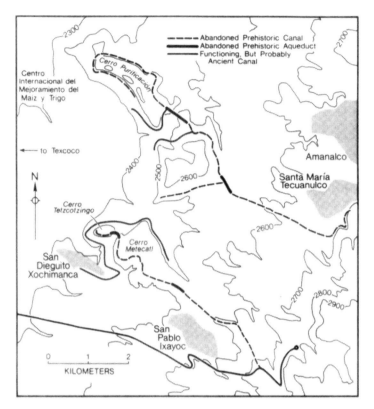

Figure 5.9. Canal and aqueduct systems in part of the Texcoco area. After Parsons 1971.

cated 7 kilometers east and a little south of the town of Texcoco, Cerro Tetzcotzingo stands out prominently at the juncture of the piedmont with the basin floor. Perhaps because of both its conspicuousness and its environmental attributes, it became a sacred place, replete with shrines and royal baths, during the reign of King Nezahualcoyotl in the fifteenth century (Niekler 1919; Mendizábal 1946). An assessment of its features found that Cerro Tetzcotzingo was intended "to represent in microcosm, the ecological components of the surrounding valley" (Townsend 1982:38). During the summer months, rainclouds formed on the peak as they do in the mountains surrounding the basin. Below the peak were springs that were used to water agricultural terraces on the slopes and fields below in a fashion analogous to the mountain streams that were used to irrigate fields in the piedmont and the valleys (Townsend 1982:61).

In spite of its cosmic qualities, including springs near its top,

Cerro Tetzcotzingo did not have a water supply of sufficient volume to maintain the few hectares of terraced gardens on its slopes. Accordingly, a canal, with aqueducts where needed, was constructed between this hill and springs high in the higher hills to the east.

Detailed studies of this canal-aqueduct system have yet to be conducted. Nevertheless, sufficient data exist from a number of sources to allow for more than a fair understanding of the technology involved (Armillas 1961:266). The only point on which any sources differ is the total length of the system, the distance from the spring to the terraces. One writer (Bribiesca Castrejón 1958:69) claimed it was 11 kilometers long; another (Dávila Bonilla 1625:619; see Wolf and Palerm 1955:271) estimated 2 leagues, or 8.4 kilometers assuming approximately 4.2 kilometers per league (Chardon 1980:137–138). This difference is obviously a minor one that might well be the result of discrepancies in estimates. All writers agree, however, that the canal was stone-lined, and one (Tylor 1861:152) noted that the blocks of stone were plastered over with a smooth stucco.

There are two points along the course of this canal where aqueducts were used to cross low points between two hills. One is north of the village of San Pablo Ixayoc, 3.5 kilometers southeast of Cerro Tetzcotzingo. In reference to this feature, one documentary source says that "mountains were levelled and valleys filled to allow the water to flow under its own power until it reached the very top of that hill" (Dávila Bonilla 1625:619; see Wolf and Palerm 1955:271–272). After its rediscovery by Pedro Armillas in the 1940s, this aqueduct was reported to be 290 meters long, 20 meters high at its highest point, and 40 meters wide at the widest (Wolf and Palerm 1955:272; see also Palerm and Wolf 1961:286). The reference to a variable width implies a trapezoidal cross-sectional shape characteristic of rock fill, thereby substantiating the claim of the earlier documentary source. More recently, however, as part of his Texcoco region settlement study, Jeffrey R. Parsons (1971:149) reported that the aqueduct was 250 meters long, 10 meters high at its center, and 2 meters wide across the top. Parsons gives no basal measurements, but a cross-sectional diagram included in his study (Parsons 1971:148, Fig. 33-a) shows this aqueduct to have nearly vertical sides.

Although the two accounts of the morphology of the feature appear to be contradictory, they are both accurate, but each reflects only a partial picture. The cross-sectional diagram provided by Parsons shows only the uppermost few meters of this aqueduct. Field reconnaissance revealed that the masonry-wall-like portion of the aqueduct reported by Parsons was actually built on top of the rock-rubble fill described by Wolf and Palerm (Fig. 5.10). In this respect,

Figure 5.10. An aqueduct near San Pablo Ixayoc involved with the conveyance of water to Cerro Tetzcotzingo. Note the approximately 2 meters of unmortared masonry atop approximately 20 meters of grass-covered rock-rubble fill. Note also the small rock fragments wedged in between the larger rocks and the multiple-layered mortared conduit on top of the masonry structure.

this aqueduct was technologically similar to the one that carried water from Chapultepec to Tenochtitlan, although the sides were steeper.

Recent inspection of this aqueduct also revealed a few other interesting points not discussed by earlier scholars. Not only did construction involve uncut rock, but small rock fragments were also used to fill the chinks. Although it is unclear whether any type of bonding agent was used to stabilize the vertical-sided upper part of this structure in order to keep it from crumbling, a mortar of rather coarse, but apparently effective, volcanic gravel was indeed used to line the conduit, that portion of the aqueduct in contact with flowing water (Fig. 5.10). Another significant, but yet poorly understood point about this aqueduct involves the conduit itself, or, more appropriately, the conduits. Parsons found a total of six levels of canal construction in this aqueduct. Apparently, it went through several stages of construction, each of which involved adding to the top of the structure. One interpretation for these rebuildings is that, being a rubble feature, the aqueduct would have settled through time. As settling occurred, the top of the aqueduct would have needed to be added to in order to connect the canals on the adjacent hills (William T. Sanders, personal communication, 1988). The only other known details about this aqueduct involve the size and shape of the conduits. Each of the six was rectangular with a flat bottom. They all measured 25 centimeters wide and 25 centimeters deep.

Further downstream from this aqueduct, and along its course toward Cerro Tetzcotzingo, the canal ran along the western side of Cerro Metecatl. There, it remains "clearly visible as an artificially built-up and levelled feature, standing up to 1 m in several places" (Parsons 1971 : 149). Accordingly, the canal here is similar to the one in the Xiquila Valley.

The canal around Cerro Metecatl terminated in a small reservoir. Details about this feature, especially its function, are lacking. All that is known is that it was stone-lined and that after departing it, water flowed across another aqueduct to Cerro Tetzcotzingo (Parsons 1971 : 123, and Plate 41).

As was the case with the aqueduct near the village of San Pablo Ixayoc, recent descriptions of this aqueduct differ from earlier ones, and for the same reasons. Describing it in the middle of the nineteenth century, Edward B. Tylor (1861 : 152) noted only that "The channel [sic] was carried, not on arches, but on a solid embankment, a hundred and fifty or two hundred feet [46–61 meters] high, and wide enough for a carriage road." Writing less than two decades ago, Parsons (1971 : 123) noted that it "is about 160 m long, and measures about 2.5 m across at the top and roughly 10 m wide at the base." He also reported that it involved "rock-rubble construction" and has "smooth, sloping walls." Most insightfully, he related that it is "built up some 7.5 m above the saddle connecting Cerro Metecatl and Cerro Tetzcotzingo." Obviously a large part of what Tylor thought was aqueduct was actually a natural landform.

At 15 centimeters the conduct carrying water across the top of the aqueduct was narrower than its counterpart farther upstream. It was, however, similar in that it was stuccoed. Indeed, the entire aqueduct was plastered over. Parsons (1971 : 123) noted that the sides were smooth. Documentary evidence lends additional support: writing ca. A.D. 1600, Fernando de Alva Ixtlilxóchitl (1891 : 2 : 210; see Wolf and Palerm 1955 : 271) described the aqueduct as "a strong and very high wall of mortar."

The terminal portion of this canal-aqueduct system involves a 500-meter length of canal flanking the south side of Cerro Tetzcotzingo (Fig. 5.11). Here, the main canal parallels the outside edge of a walkway some 2 to 3 meters in width, partially cut into the hillside, and partially built up and reinforced with rock walls (Parsons 1971 : 123). At least some of the water from this stucco-lined canal was used to fill two features identified as royal baths, remains of which are still

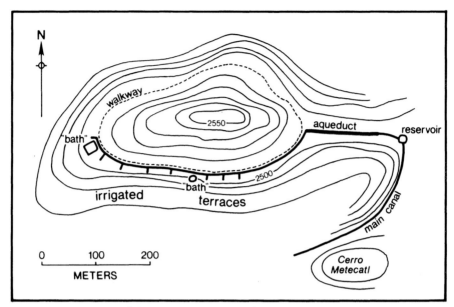

Figure 5.11. Map of Cerro Tetzcotzingo. After Parsons 1971.

found along its course. Most, however, was diverted out at right angles, through rock troughs, and used to irrigate terraces on the slopes below (Parsons 1971: Plates 42-b and 43-a). Today, most of these terraces are abandoned because water no longer flows through the system. Some of the lower terraces immediately above the present-day village of San Dieguito Xochimanca (Fig. 5.9) are, however, irrigated with water brought through a canal from a spring on the opposite side of Cerro Tetzcotzingo (Parsons 1971: 121).

The other canal-aqueduct system that provides insight into the technological developments involving late prehistoric canal irrigation in Mexico is that associated with nearby Cerro Purificación, located 7 kilometers due east and a little north of Texcoco (McAfee and Barlow 1946: 113). This system is, in many ways, similar to that associated with Cerro Tetzcotzingo: it brought water from springs several kilometers away; the inside dimensions of the canal and conduits were rather small, averaging 25 centimeters deep and 25 to 50 centimeters wide; it involved two aqueducts; and in places the canals were excavated into the hillside, while in others retaining walls were built to support it on the slopes. On these last two points, the system also differed substantially from its neighboring counterpart.

The aqueducts associated with Cerro Purificación are significantly larger than those previously discussed. Located 3 kilometers due west of the village of Santa María Tecuanulco (Fig. 5.9), one of these aqueducts measures 375 meters long, between 2 and 5 meters high, and 2 meters wide at the top (Parsons 1971: 147).

Even larger, however, is the aqueduct connected directly to Cerro Purificación. Measuring 13 meters high at its midpoint, 2 to 3 meters wide across the top (Parsons 1971 : 147), 13 meters wide at the bottom, and 1,010 meters in length (Wolf and Palerm 1955 : 269), this aqueduct is clearly the largest one in the Texcoco area. As with the others, this aqueduct involved rock-rubble construction. Unlike other aqueducts that had bases twice as wide as the total heights of their structures, this one had a base that equaled the height. Given that the top was up to 3 meters in width, this aqueduct is truly wall-like and has nearly vertical sides (Fig. 5.12).

Construction of vertical walls was, of course, no problem to architects and engineers in the Valley of Mexico during Aztec times. Indeed, structures with similar characteristics were found throughout the large cities of the day. What is different here, however, is that the "walls" were used not in buildings, but for water conveyance. Unlike buildings that have cut, fitted, and mortared masonry walls at right angles to—and therefore supporting—each other, single-walled rock-rubble aqueducts that are free-standing must be engineered with great precision. That these structures have stood for more than five centuries is indicative that their builders had achieved a very high level of expertise.

Figure 5.12. View along the top of the aqueduct at Cerro Purificación. Note the rock-rubble construction and the steep sides.

Supporting this conclusion is evidence involving the second point
on which the Cerro Purificación aqueduct differs from others in the
Texcoco area. Further downstream from this feature, and around the
sides of the hill, the canal carrying water to terraces along the slope
is in some places cut into solid rock (like the canal at Monte Albán
Xoxocotlan), while in others it is built up with rock walls (like the
Xiquila canal). What is different here is the size of the built-up portions and the manner in which they are supported. In many places
along the slope, the canal is built up to heights of 1 to 3 meters (Parsons 1971: Plate 57-b). In a few others, however, the walls are up to
10 meters high and supported by two- and three-tiered buttresses
(Parsons 1971: 147). Features of this size and character are known
from no other place in Mexico. Furthermore, they not only demonstrate a high degree of engineering proficiency, but also indicate that
their builders were on the verge of making a major technological
breakthrough.

As great an accomplishment as the construction of a wall-like
aqueduct was, its contribution to the future development of canal
irrigation technology was thwarted by the Spaniards. Had they not
arrived and introduced arched masonry aqueducts, the indigenous
irrigators in Mexico probably would have developed the technology
themselves within a century or so. Evidence for this conclusion is
apparent both in the developments that were taking place and in the
remains of rudimentary arched structures in the vicinity.

The construction of wall-like aqueducts, when seen in light of the
aqueducts that carried water across the western part of Lake Texcoco from Chapultepec Hill to the Aztec capital of Tenochtitlan,
was in all probability the step immediately preceding the development of arched masonry aqueducts. The aqueduct over the lake was
certainly a masonry and bridge-like structure with openings beneath. It is not known if any of the aqueducts in the Texcoco area
had any openings or drains running through their bases. It really
makes no difference, however. Earlier dams had openings at their
bases in the form of floodgates. It is logical, therefore, to conclude
that all the necessary pieces of technology were available and ready
to be incorporated. Supportive evidence for the impending development of arched masonry aqueducts is also found in the existence of
what has been claimed to be a pre-Hispanic arched masonry bridge
in the Texcoco area.

In addition to providing information on both the canal and the
aqueduct that carried water to the top of Cerro Tetzcotzingo, Tylor
(1861: 153–154) also reported an ancient bridge near the point where
Cortés launched his brigantines during the Conquest. According to

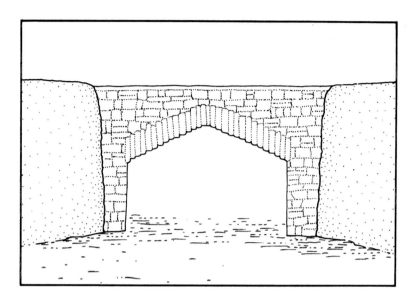

Figure 5.13. Pre-Hispanic arched bridge built near Texcoco. After Tylor 1861.

Tylor, the underside of the bridge is made in the form of a gable rest-
ing on two buttresses nearly 7 meters apart (Fig. 5.13). It was composed of slabs of stone placed on end and topped by stones laid hori-
zontally. Although the stones were mortared together, the vertical
ones were sufficiently irregular in shape to support the structure.
While this bridge is an interesting feature in many regards, its most
important characteristic, assuming that it is pre-Hispanic, is that it
shows "how nearly the Mexicans had arrived at the idea of the arch"
(Tylor 1861:164). That they soon would have applied this technology to the construction of aqueducts and, indeed, improved it further during the process seems inevitable.

6. Origins and Cultural Implications

The discovery of monumental hydraulic features, such as the aqueducts constructed during the late pre-Hispanic times, has undoubtedly precipitated much of the scholarly attention that has been given to prehistoric canal irrigation in Mexico. Although much has been learned about these and lesser features, especially during the past three decades, at least two questions that inherently involve technology remain unanswered. One deals with the origins of canal irrigation; the other is concerned with its cultural implications. At one time or another, probably everyone interested in canals has thought about the very first irrigator and how the idea of constructing an artificial watercourse came about and was subsequently implemented. There also have been numerous discussions, many quite lengthy, and probably tens, if not hundreds, of thousands of printed pages devoted to understanding the social, political, and economic organization of civilization based on canal irrigation. This chapter addresses both of these topics.

Possible Canal Origins

Over the course of three millennia, the prehistoric farmers of the region known today as Mexico made considerable advances in the development of canal irrigation technology. Exactly how the idea of watering crop plants by means of artificial watercourses came about and how it began to be expanded upon, however, long have been a mystery (Woodbury and Neely 1972:127; García Cook 1985:61). After suggesting that canal irrigation might have had more than one place of origin in Mexico, Palerm (1955:35−36) helped to initiate systematic research on the topic by stating that "The problem of dating the beginning of irrigation in Mesoamerica can be solved only by archaeological means." Wolf (1959:77) not only supported Palerm's point but reiterated it, claiming that "Irrigation may have been practiced early. . . . But again we shall be merely guessing about its presence . . . until we have more certain data."

A great deal of in-depth research certainly has been conducted since Palerm and Wolf made their initiating comments. That we are getting closer to understanding what actually took place early on is clear. It is equally evident, however, that much more work needs to be done.

The earliest confirmed date for the use of irrigation canals in Mexico is ca. 1000 B.C. Although this might seem to be quite early, there have been suggestions that the practice is much older, dating back perhaps as early as the origins of agriculture (Flannery et al. 1967:453). Writing seventy-five years ago, Herbert J. Spinden (1915: 270) first proffered that "Irrigation is often looked on as a remarkable sequel of the introduction of agriculture into an arid country. But from the best historical evidence at our command we should rather regard it as an invention which accounts for the very origin of agriculture itself" (see also Spinden 1917:182; 1928:52–53). Exactly what Spinden based this assertion on is unknown. Quite literally, the only other thing he had to say on the topic was that "we have every reason to believe that the earliest agriculture was developed under irrigation" (Spinden 1928:63).

If this assessment is correct, and given that the earliest confirmed evidence of domesticated crops in Mexico dates to ca. 5200 B.C. (see Mangelsdorf 1974:165), a gap of four millennia exists in the developmental sequence of the canal irrigation technology. Because the known sequence now spans 2,700 years, the unknown events that occurred prior to 1000 B.C. constitute at least 60 percent of the entire history of canal irrigation in Mexico. Although confirmed canal data are currently lacking, researchers have not been hesitant to speculate on what took place during these early times in the development of the technology. Essentially, there are three schools of thought on the issue. One school suggests that the first canals were developed to carry water from small, permanently flowing springs to agricultural plots, specifically in the Valley of Mexico (William T. Sanders, personal communication, 1988). Unfortunately, this idea has not been stated in print, and the rudiments of the scenario have not been outlined. Furthermore, and more important, there currently exist no data to support such a notion. Given these conditions, further discussion of the matter is neither warranted nor justified. Such is not the case, however, for the two other schools of thought.

The first of these maintains that the earliest canals used for irrigation purposes originated as drainage ditches that carried water away from habitation sites (e.g., Flannery and Marcus 1976:378). Data from the Teopantecuanitlan site tend to support this position. The second of these schools holds that the use of canals for irrigating fields evolved out of floodwater farming, the practice of planting in areas watered by excessive natural stream flow (e.g., C. E. Smith 1965:77). Although speculative, this idea merits discussion because evidence exists for the very early use of floodwaters for irrigation.

The Drainage Theory

The use of drains is an elementary technological development that might well have been the precursor of canal irrigation. Carrying water away from an area is technologically much easier than transporting water to a specific place. For drainage purposes, a ditch can be excavated almost anywhere as long as it is lower than, and points in any direction away from, the place to be drained. Attention to hydrologic characteristics can be minimal because the nature of the flow and the destination of the water are not necessarily important. In contrast, getting water to a specific place, as is the case with irrigation, is problematical. The canal must not only be lower than the source of the water, it must also be higher than the field. Furthermore, it has to follow a route that has neither too steep nor slight a gradient. Care must be taken to minimize downcutting, sedimentation, meandering, and overflow.

Because of the differences in their respective technological complexities, it seems plausible that drainage features should predate irrigation canals. Paleolithic people, for instance, were certainly capable of digging small ditches in order to drain water from cave floors millennia before crops were domesticated. In a less extreme example, drains not unlike the canal postulated for the site of Teopantecuanitlan were used quite early at the famous Olmec site of San Lorenzo in present-day Veracruz state (Coe 1968:41–71).

If the canal at Teopantecuanitlan stands the test of time and is eventually accepted as the earliest irrigation canal in Mexico, then evidence will exist to substantiate that canal irrigation could have developed out of site drainage. The buildings at the site, and presumably the drains, predate the currently suspected irrigation features by two hundred years. Assuming that the inhabitants of the site, like prehistoric agriculturalists almost everywhere, cultivated small plots near their homes (e.g., Anderson 1954:136–151), drainage from the site would have contributed to their watering. The stone-lined drains at Teopantecuanitlan, therefore, might well have been the prototypes for the canals found at the site.

Along similar lines, the rising level of water inadvertently impounded in the plaza as the result of the surrounding structures could have provided farmers insight into the building of a dam and construction of a reservoir. In effect, the development of a reservoir by constructing a dam across an ephemeral drainage is similar in some respects to creating a flood-prone central plaza by enclosing an area with several masonry structures. Technology involving the impoundment and diversion of water for irrigation could well have been developed out of the need to solve a totally unrelated flooding and drainage problem.

The scenario by which the origin of canal irrigation is envisioned as having developed out of the drainage of habitation areas is certainly plausible. It is also meritorious in that it can be used to explain the technological discongruity involving the existence of a seemingly advanced rock dam and stone-lined canal so early at Teopantecuanitlan. As will be recalled from Chapter 2, dams of any kind, but especially rock ones, do not appear elsewhere in Mexico until several centuries later than that purported from this site. Although it is tempting to conclude that the dam at Teopantecuanitlan and the canal as well were major technological innovations well ahead of their time, such is probably not the case. The site itself involved masonry structures. Building a dam, and lining canals with the same materials, not only would have been a logical and non-revolutionary technique, but might well have been the only one conceivable. Indeed, it is highly unlikely that those who constructed the masonry structures at the site would use anything other than rock to build a dam and a canal, especially if the latter was linked to stone-lined drains.

Exactly why a hiatus occurred in the construction of dams and why canal technology regressed remain unknown. In all likelihood, however, subsequent irrigators who acquired knowledge of diverting water and transporting it through canals, from their mentors or predecessors at Teopantecuanitlan, either discovered quite quickly or learned in a reasonably brief period of time that rock features were not necessary. Moreover, rock dams and canals both required some degree of masonry expertise and a great deal of labor. If this argument is correct, then the technology of building dams and stone-lined canals was probably lost. Dams were not rediscovered until a few hundred years later in the Tehuacan Valley, and building them out of rock took another two centuries when one was constructed at Monte Albán Xoxocotlan in Oaxaca. Both of these features were parts of systems that involved ephemeral water sources—runoff or floodwaters.

The Floodwater Theory

Ideas concerning the development of canal irrigation out of flood-water farming can be traced directly to classification schemes used to describe the different types of water control systems for which archaeological evidence has been found. Once researchers had categorized the various systems they found, it was a rather easy step for them to determine differences in technological complexity as evident by the degree of difficulty involved, or expertise needed. For example, Anne V. T. Kirkby (1973:36) noted that "in their simplest forms floodwater techniques merge with those of dry farming. At the other end of the scale, floodwater distribution systems are similar to those of canal irrigation."

Determinations such as this typically were made on the basis of present-day traditional or unmechanized agricultural activities as observed by researchers: ethnographic analogs have been long used in attempts to understand prehistoric agriculture. The sequences in which various systems developed were inferred from differences in technological complexity. Simple systems were presumed to be earlier than relatively more complex ones.

A number of different classification-development schemes have been outlined (e.g., Woodbury 1962:302; Palerm 1967:37). Although each has its strong points, all are highly generalized and none covers the technological intricacies of the various systems' components. Elements of each scheme, however, can be extracted and pieced together in order to form a rather detailed picture of how floodwater farming developed and contributed to the development of canal irrigation coevally with domestication.

Virtually everyone working on the topic of the origins of agriculture accepts that, however they were evolving or being domesticated (see Crosswhite 1981–1982), the plants that would eventually become economically important staples were first collected and nurtured in small, naturally well watered plots where minimal amounts of labor and technological expertise were needed (e.g., Rindos 1984: 82–144). In arid areas, where canal irrigation eventually would be necessary for production on anything other than a very small scale, the first fields are thought by some (e.g., Woodbury and Neely 1972: 150) to have been located on alluvial deposits where channels had not been deeply incised. Such locales are watered by periodic stream flow originating either as convective rainfall elsewhere in the drainage basin or as orographic rainfall in the higher elevations. It has also been suggested that these areas would have been attractive to early farmers because the periodic but often torrential surface flows would have resulted in the deposition of fresh fertile sediment, and they would have facilitated clearing by impeding the growth of natural vegetation (C. E. Smith 1965:77).

Getting water onto cultivated areas was certainly no problem for these early farmers. Controlling the water was, however, most problematical. Presumably, ancient cultivators who relied on floodwaters, like their counterparts in many parts of the world today (e.g., Nabhan 1986b), planted only after the soil had been sufficiently wetted by the first or second flood. Under normal conditions this would be all the surface water that was needed. Alluvium in ephemeral tributaries is often quite coarse, and water in the form of subsurface flow persists long after the surface flow disappears (C. E. Smith 1965:77). This subsurface water would have been used to sustain the crop throughout the growing season (e.g., Nabhan 1986a:49). The greatest problem these farmers faced was to insure that later floods would not wipe out their crop.

A number of devices of unknown origin and antiquity have been developed in order to minimize the impact of floodwaters on low-lying cultivated areas. The simplest of these, and presumably therefore the most ancient, involves the piling of brush and rocks around the upstream ends and channel sides of fields (e.g., Doolittle 1984a: 128). Technologically more complex water spreaders were constructed, presumably sometime later, by anchoring brush with stakes driven into the ground. Instead of being large and enclosing much of the cultivated area, water spreaders are typically small and strategically located throughout the fields (e.g., Nabhan 1979:249). Rather than attempting to keep water off the fields, these devices quite literally, as their name implies, spread surface flow, thereby slowing its erosive velocity and minimizing the formation of gullies.

Sheet flooding of ephemeral streams often cuts small gullies that remove not only vegetation, including crops where cultivated, but also soil. Because flowing water tends to follow a path of least resistance, gullies usually head at the outside edge of channel meanders and take the most direct course downstream: they flow longitudinally, straight down the floodplain (e.g., Richards 1982:180–220). Although gullies must have had a negative impact on ancient cultivation, they also were advantageous, especially to incipient irrigators. Gullies may well have provided observant farmers insight into the mechanics of redirecting the flow of water.

Early on farmers undoubtedly cultivated plants in areas that received water naturally by means of such gullies. Although they probably did not realize it at first, these farmers were "irrigating" their crops inadvertently, using gullies much as their later counterparts used canals. Presumably, at some later time, inventive cultivators perceived the hydraulic functions of the gullies and began to mimic them by intentionally excavating small canals that turned out to be prototypes for later systems. The transition from using and observing natural gullies to excavating canals in places where gullies did not exist was technologically quite simple. Once mastered, however, the science of canal construction could be applied to those places where flooding and gullying were not common occurrences and where water deficiencies would have made canal irrigation not only an attractive alternative, but one essential for cultivation (Woodbury 1962 : 302).

The development of stake and brush water spreaders to control gullying also probably contributed to the development of canal irrigation technology. Although they differ somewhat in size and function, water spreaders are morphologically similar to weirs. That these small gully-prevention features were the prototypes of larger water diversion structures seems plausible. Indeed, water-spreader-type features might well have been tried and used to control erosion at the mouths of canals sometime after they had been perfected on the fields and after the rudiments of canal construction had been worked out.

The transformation from water spreaders on fields, through erosion control devices at the bifurcations of gully-canals and channels, to full-blown weirs was a small but important technological advancement (Flannery 1983 : 333). Although it involved little more than the use of existing devices to remedy a new problem, the development of weirs meant that canals could be built and function in places where they could not have done so previously. The development of weirs also played an important role in the development of diversion dams and, eventually, storage dams.

Given their temporary nature, there is no way in which weirs can divert substantial amounts of water to irrigate large tracts distant from stream channels. In order for less flood-prone lands to be irrigated, early farmers would have had to excavate longer canals, and build more permanent and effective diversion devices (Palerm 1967 : 37). Doing the latter would have involved nothing more than piling more rocks onto the brush features across stream channels; it also would have resulted in some water being retained upstream of the devices well after the streams stopped flowing. The inadvertent retention of waters, while initially neither planned nor desired, would soon have been seen as having some clear advantages. Most specifically, the water retained could be used for irrigating fields during critical times of the growing season when rainfall deficiencies often exist. At some later time, dams were built deliberately in order to impound water for storage purposes. Although it is unknown when specific developments took place, tentative evidence from Teopantecuanitlan suggests that dams were used by 1200 B.C.

Technological Development and Social Organization

During its earliest years, canal irrigation technology must have changed very little. Accomplishments were undoubtedly few and small, as little survives in the way of material remains. Although the dam, canal, and associated drainage features at Teopantecuanitlan were parts of what was a reasonably intricate system for its time, the fact remains that technological achievements were not very great prior to approximately 600–500 B.C. Indeed, the only other confirmed canals from early times are the simple floorwater irrigation canals from Santa Clara Coatitlan and the Meseta Poblana— and if the sequence just outlined for canal irrigation having originated with floodwater farming is correct, these canals would have been rather minor technological developments. It was only after irrigation began to be practiced near the site of Monte Albán in Oaxaca that substantial developments started taking place.

The improvements that occurred in canal irrigation technology around 550 B.C. were especially notable (Fig. 6.1). Not only were storage dams constructed of uncut, loosely piled rocks, but, in part at least, they were also being built of cut blocks mortared together. Spillways were improved in some cases, while in others floodgates were built. Unlike earlier dams that can be classified as gravity dams, the later ones were arch dams. Although the evidence for their use is scant, it also appears that rudimentary diversion dams must have developed ca. 300 B.C. Such features, along with head gates, would have been needed in order for water to be diverted out of perennial streams.

In addition to water diversion structures, the canals themselves underwent rapid modification and elaboration at this time. Whereas the earliest suspected canal was lined with stone slabs, most—and certainly all those confirmed to date—were unlined and earthen. Canals also had numerous cross-sectional shapes. Some canals were rectangular, some were V-shaped, others trapezoidal, and yet others U-shaped. A few canal systems exhibited a great deal of technological knowledge on the part of their builders. The one at Monte Albán Xoxocotlan, for example, was both "stepped" in cross-sectional

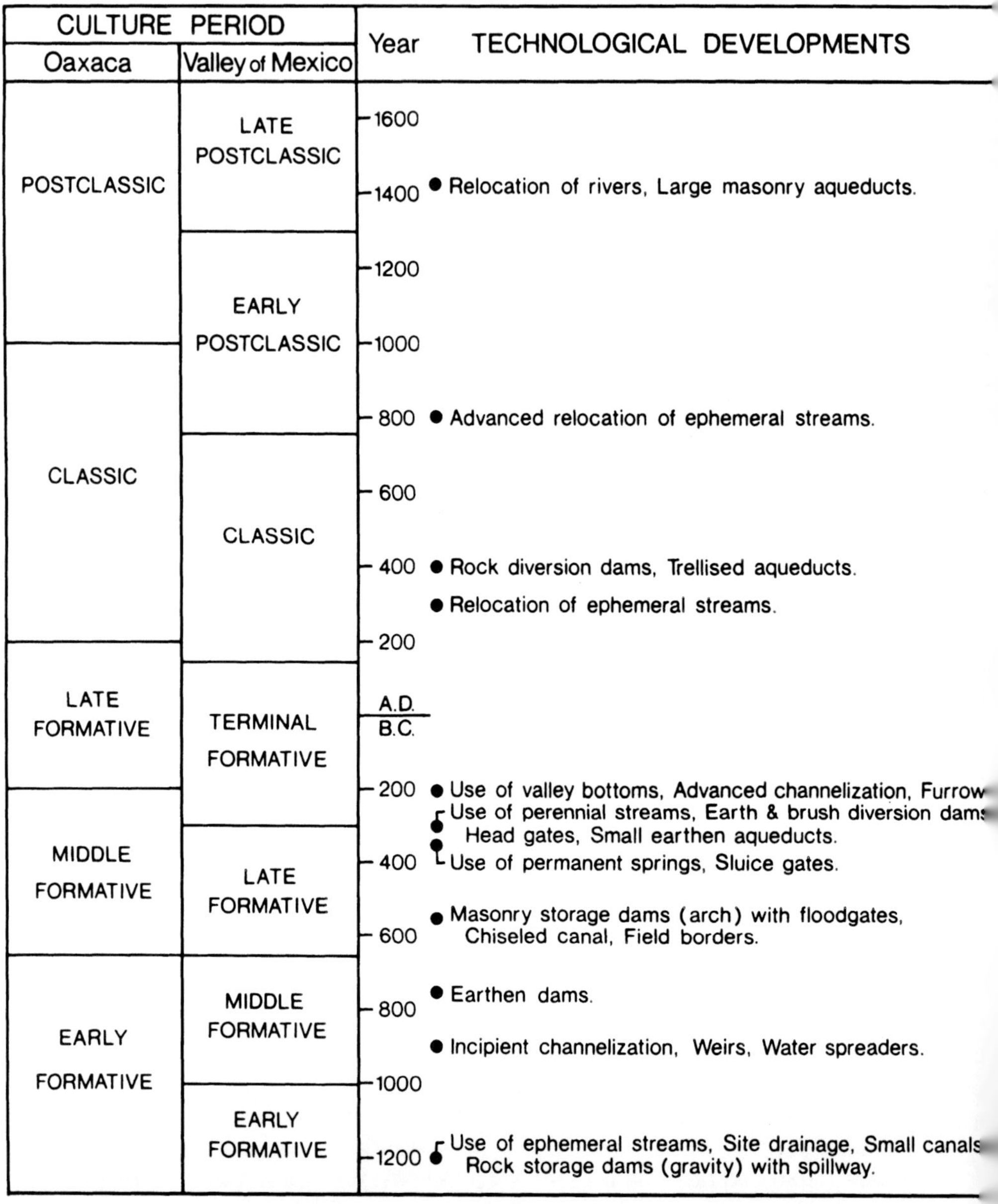

CULTURE PERIOD		Year	TECHNOLOGICAL DEVELOPMENTS
Oaxaca	Valley of Mexico		
POSTCLASSIC	LATE POSTCLASSIC	─1600	
		─1400	● Relocation of rivers, Large masonry aqueducts.
	EARLY POSTCLASSIC	─1200	
		─1000	
CLASSIC	CLASSIC	─ 800	● Advanced relocation of ephemeral streams.
		─ 600	
		─ 400	● Rock diversion dams, Trellised aqueducts.
			● Relocation of ephemeral streams.
		─ 200	
LATE FORMATIVE	TERMINAL FORMATIVE	A.D. B.C.	
		─ 200	● Use of valley bottoms, Advanced channelization, Furrow
MIDDLE FORMATIVE	LATE FORMATIVE		┌ Use of perennial streams, Earth & brush diversion dams ● Head gates, Small earthen aqueducts.
		─ 400	└ Use of permanent springs, Sluice gates.
		─ 600	● Masonry storage dams (arch) with floodgates, Chiseled canal, Field borders.
EARLY FORMATIVE	MIDDLE FORMATIVE	─ 800	● Earthen dams.
			● Incipient channelization, Weirs, Water spreaders.
		─1000	
	EARLY FORMATIVE	─1200	┌ Use of ephemeral streams, Site drainage, Small canals ● Rock storage dams (gravity) with spillway.

Figure 6.1. Regional chronology and dates of developments in various aspects of canal irrigation technology.

shape and carved into bedrock as well as being excavated in loose earth. Also, the one at Loma de la Coyotera had small aqueducts that facilitated the transport of water over low-lying areas.

On the fields themselves, water was no longer applied only by means of wild flooding, but bordered checks were built around and within some fields, and crops were irrigated with water that was carefully controlled.

Between 550 and 200 B.C., developments in both irrigation-related features and entire canal systems were numerous and varied. Improvements in different types of features allowed for several environmental settings to be irrigated. Ephemeral streams, springs, and even perennial rivers were being tapped for irrigation waters. Furrows also provided for better control of water on the fields than had border checks developed previously. Perhaps the greatest development of the time, however, involved the channelization of stream beds in conjunction with the excavation of canals and the construction of dams to control water at Amalucan, near present-day Puebla city. These developments facilitated the cultivation of broad expanses of valley-bottom floodplain land.

In sum, much had been achieved in canal irrigation technology in a very brief period. Most of these developments occurred during the Middle Formative period (600–200 B.C.) in and around what is today Oaxaca. That such achievements were made is remarkable in itself. That they took place in such a short period of time is even more so. Presumably, because they came about at the end of a long time span during which there is little evidence of canal irrigation but abundant reason to believe it was being practiced in some rudimentary form, these accomplishments were the culmination of a great deal of experimentation. In effect, these achievements reflect a developmental sequence that can be best described as having occurred exponentially. A long infancy was followed by a brief period in which prior technological attempts crystallized into a myriad of well-developed canal irrigation systems.

Paradoxically, canal technology stopped developing as abruptly as it gelled. After 200 B.C. virtually no new developments occurred for approximately five hundred years (Fig. 6.1) It was not until around A.D. 300 that anything new came about, and then developments occurred sporadically for over a millennium. The first notable development was the channelization and relocation of an ephemeral stream on the Tlajinga Plain. A century later, around A.D. 400, a rock diversion dam and trellis-like aqueducts over deep ravines were built in the Xiquila Valley. It was another four hundred years before a third, and yet another six hundred years before a fourth significant tech-

nological change occurred. These developments involved the relocation of a stream channel to carry water to the Maravilla canal around
A.D. 800, and the monumental works of the immediate pre-Hispanic
era, respectively.

Starting in the Late Formative Period in Oaxaca and the Terminal
Formative Period farther north in the Basin of Mexico (ca. 200 B.C.),
canal irrigation technology changed very little. With the exception
of the first two developments just noted, the technology remained
essentially the same through the Classic (A.D. 200–800/1000) and
Early Postclassic (A.D. 800/1000–1300) periods. Stated another way,
canal irrigation as it is now known was developed largely before that
period in which prehistoric people in Mexico attained a sufficient
level of social, political, economic, and artistic complexity to be
considered a state.

On the basis of this finding, it might appear that those scholars
who have argued that water control in general and irrigation specifically not only preceded but precipitated formation of states are correct (e.g., Wittfogel 1957). On closer inspection, however, it is quite
clear that the problem is much more complex (Price 1971) and certainly due more consideration than is allowable within the scope of
this work. Suffice it to say for now that virtually all of the arguments
concerning the relationship between water control and ancient states
involve large-scale systems covering thousands of hectares (R. McC.
Adams 1968:370–371; Robert C. Hunt, personal communication,
1988), and few of the canal irrigation networks known to have existed in Mexico prior to late pre-Hispanic or Aztec times can be so
considered (Table 6.1). Although some, such as the valley-bottom
system that involved travertine-encrusted canals on the Llano de la
Taza, eventually involved some reasonably expansive tracts of land
and were technologically complex, they were not on the same order
as those in Huang Ho Valley, for example, and they were small and
elementary in terms of their sociopolitical implications.

For instance, in discussing the manner in which those canals were
constructed, Woodbury and Neely (1972:127) stated that "consideration of the details of field systems and evaluation of the actual
work involved provides a basis for viewing these field patterns as the
expectable results of moderate efforts and not as fantastic achievements of the days of the giants." Referring to the scale of operation,
they noted that "it would be wrong to imagine that there is evidence
for a sophisticated, widespread system of water management. Each
field system is not only a small, self-contained unit, but it is probably the work of a single farmer or family group" (Woodbury and
Neely 1972:125).

Table 6.1. *Sizes of individual canal systems in prehistoric Mexico as measured in terms of maximum possible area initially irrigated*

Site	Hectares
Teopantecuanitlan	few
Santa Clara Coatitlan	10–20
Meseta Poblana	35
Monte Albán Xoxocotlan	50
Santo Domingo Tomaltepec	50–60
Amalucan	70
Hierve el Agua	50
Loma de la Coyotera	75
	(eventually expanded to 737)
Otumba	few
	(eventually expanded to 200)
San Buenaventura	50
Llano de la Taza	130
	(eventually expanded to 1,300)
Cerro Gordo	50
Tlajinga Plain	15–20
Tecorral Canyon	1
Xiquila Aqueduct	20
Maravilla	100
Trincheras Area	35
La Quemada	50
Chalchihuites	few (?)
Tula	200
Eastern Sonora	150
Conchos	few
Casas Grandes	1,000
Zempoala	few (?)
Basin of Mexico (in general)	few hundred each
Chapultepec	500–1,000
Cuautitlan	1,000–2,000
Texcoco Area (Aztec times)	several hundred to thousands

Although the technology did not change significantly, and the systems were for the most part small, not involving a great deal of land, it should not be assumed that canal irrigation was no longer important or that it was not being practiced in locales where it had not existed previously. The contrary is actually true. During the Classic Period canal irrigation was more widespread than at any earlier time.

The reason for this apparent paradox lies in large part with the physical environs of the region. There exist numerous locales suitable for irrigation but none with sufficient amounts of surface water for the operation of large-scale systems (Rojas Rabiela 1985:192). This environmental-technological relationship has been no better stated than by Wolf, who noted that

> Irrigation has been important in Middle America; everywhere
> canals and small dams perform the task of storing and
> channeling water to dry fields, to secure the first crop against dry
> spells, to help grow the second crop on otherwise sterile land.
> But the very lay of the land inhibited the growth of large-scale
> irrigation, and thus of the all-dominant, overweening hydraulic
> state. There are no great bodies of water, only a rivulet here and
> there. Except in the lowlands where irrigation is not needed,
> there are no great, permanent, slow-moving rivers that could be
> held back or diverted easily through massed effort; all too often
> streams of water tumble precipitously from the steep mountains,
> too fast to be tapped except in occasional feeder canals. There are
> no great wide-open spaces that could be saturated with water
> from one canal; too often mountains protrude to shut off one
> irrigated valley from its neighbors. Everywhere the patches of
> irrigated land are visible, like cool green islands against a
> background of drab and arid soil, but to this day irrigation has
> remained essentially localized, insular, in a world of uncertain
> rainfall (Wolf 1959:16).

The nearest thing to an exception to this scenario is the Central Mexican Plateau, including southern Hidalgo, Tlaxcala, western Puebla, and Morelos states, where, at contact the Spaniards reported, but did not describe or discuss, large irrigation systems (Sanders 1956). In the Basin of Mexico, where good archaeological data about prehistoric irrigation do exist, broad expanses of valley-bottom land are relatively abundant, and surface waters are close to the magnitude required according to the literature on hydraulic civilizations. Numerous streams head in the surrounding mountains and empty into the basin-floor lakes. The channels of these streams dissect the alluvium, thereby creating several large tracts that can be irrigated most adequately by series of canal networks that carry water diverted out of the streams.

Although many of these lands are no longer being cultivated because of present-day urban encroachment, most, if not all, were irrigated in prehistoric times. During the Classic Period in the basin, many of the lands in the northern part of the region, and certainly all of those in the Teotihuacan Valley, were being irrigated, but only by means of reasonably small floodwater and permanent canal networks (Sanders 1981 : 180–182) that, contrary to what some scholars (e.g., Matheny and Gurr 1983 : 81) think, did not require highly organized labor for construction and maintenance. The irrigated area in the Basin of Mexico appears to have decreased somewhat during the Early Postclassic Period (Sanders 1981 : 183–185). By the Late Postclassic Period, however, every available tract of land was most likely being irrigated intensively, and virtually every type of water control system known was being employed (Sanders 1981 : 189–191). Although most remained small, watering only tens or hundreds of hectares, with technology that had been long perfected, a few reasonably large canal networks irrigating several hundred and, indeed, a few thousand, hectares were developed at the time (Table 6.1).

In at least one case, the Río Cuautitlan channelization project, water was diverted out of one river channel, across the interfluve, and into a second valley. This interconnected valley system is the closest that ancient Mexicans came to developing a canal network comparable in size, scope, and technological complexity to those where hydraulic civilizations arose in the Old World. Although it pales in comparison to some of the irrigation systems developed in such places as Mesopotamia, the Cuautitlan system did involve engineering expertise and coordinated group labor on an order not known previously in Mexico. That this development occurred so late in prehistoric times puts a kink in the argument that large-scale water control, and specifically irrigation, gave rise to ancient states. It does not, however, support the opposite argument that large-scale irrigation systems were developed only by societies already possessing high degrees of social organization (e.g., Kappel 1974 : 163). Instead, it appears that major developments in canal irrigation technology developed during, and coevally with, the rise of the Aztec state. Evidence indicates that the truly monumental irrigation works known prehistorically in the basin were all built at a time when the population was near its zenith—a phenomenal 1.2 million with a

density of 180 per square kilometer (Whitmore and Turner 1987)—
and while social, political, and economic institutions were operating on the scale of a fully theocratic state (Porter Weaver 1972).

Debate on the cultural implications of water control, and specifically irrigation, is far from over (e.g., B. de Lameiras 1986; R. Hunt
1988; Rojas-Rabiela 1988 : 123–154), and, of course, this assessment
of canal technology and its development was not intended to settle
the dispute. It does, however, reveal something about the nature of
the relationship, especially how it changed through time. Canal irrigation networks, regardless of how technologically complex or the
amount of land involved, are, by their very nature, hydrologic systems. In order to sufficiently irrigate a given parcel of land, a canal
has to have a certain discharge (measured in cubic meters per second), which is a function of the cross-sectional area of the canal
(measured in square meters) multiplied by the velocity of the flowing water (measured in meters per second). Velocity is dependent in
part on friction, a function of both roughness of the canal surface
and the size of the wetted perimeter (the distance from the surface of
the water down one side of the canal, across the bottom, and up the
other side, measured in linear meters). Because canals carry water by
means of gravity, velocity is also dependent on the gradient, which
itself is in part dependent on the sinuosity of the canal. Put simply,
canals that are smooth, wide, shallow, steep, and straight have higher
velocities, and hence greater discharges, than ones that are rough,
deep, narrow, nearly flat, and winding.

Making sure that these factors are in perfect harmony is important for irrigators; otherwise the system will not function properly.
For example, a canal that is too steep might carry water at an excessive velocity, thereby promoting lateral erosion along the curves and
downcutting throughout its entire length. A canal that is too sinu-
ous, on the other hand, might flow too slowly and suffer sedimentation problems as a result.

Of course, no canal irrigation system is going to function perfectly.
By its very nature, flowing water creates perturbations in the system
that irrigators have to deal with. These vary considerably, but there
is always something that needs to be done. For example, picture the
consequences that could befall an irrigator who removes sediment
that has been accumulating in one part of a canal. Immediately upon
removal of the sediment, the velocity of the flow will increase. This
increase could then result in downcutting, lateral erosion of canal
banks, especially on the outsides of curves, or both. How could this
erosion problem be resolved? One way would be by rip-rapping the
canal bank where it is eroding; another would be by lining the canal.

Neither of these methods, however, will slow the flow. Indeed, the latter might even increase it more. One way of diminishing the velocity would be by constructing weirs in the canal (International Land Development Consultants 1981 : 227–230). These would slow the water, but they would also trap sediment, thereby recreating the initial problem.

In sum, canal irrigation systems are not static features on the landscape: they are constantly changing. Irrigators are, quite literally, locked in an eternal struggle, battling the effects of perturbations created by the solutions to previous problems.

In many cases, changes that take place in the normal course of cultivation and maintenance result in what has been identified as "incremental" expansion of the system (Doolittle 1984a). Under such circumstances, new fields are created by gradually clearing, and constructing features such as terraces and field borders, over a number of years. Paralleling the sequential additions is the expansion of the canal network. Although this manner of development might seem obvious, it is often assumed that canals are expanded simply by adding new sections to their downstream ends (e.g., Nicholas and Neitzel 1984 : 163). There might be some cases in which this type of expansion is possible, but they are certainly unusual. For the most part, once canal systems are in place and operating, they can be expanded only by rebuilding the entire existing network. Such rebuilding typically involves technological changes and has immense cultural implications, especially where very large systems are involved.

Assuming that all of the water flowing through a canal is used, a functioning canal—with a given gradient, cross-sectional area, and shape—can carry only so much water, and, therefore, irrigate only so much land (Zimmerman 1966 : 203–234). The excavation of a second canal, intended to take water out of the first, will not normally increase the total size of the irrigated area. The second canal will result in some previously unused land being irrigated for the first time, but it will also result in water shortages on lands previously cultivated. The only way that such water shortages can be ameliorated is by taking part of the originally cultivated land out of production. The end result of simply adding a second canal, therefore, does not necessarily include an increase in the irrigated hectarage.

If cultivators wish to irrigate larger amounts of land, they usually need to do more than simply add new canals to their existing network or extend the canals currently in use. They also have to enlarge those segments of the canals upstream of the points where the new additions bifurcate or new sections are added. Enlarging canals is not, however, a simple and straightforward process, as it results in changes in the cross-sectional areas of a canal (A), its wetted perimeter (P), and, hence, the hydraulic radius (R), which equals A/P (see, e.g., Israelson and Hansen 1962:81). Changes in this last dimension are especially critical, as its square root, assuming other variables (e.g., roughness, sinuosity, and gradient) are held constant, is proportional to the velocity of flow (V) (see, e.g., Morisawa 1968:35), which, when multiplied by A, equals discharge (Q). For example, a canal that is 1 meter wide, 1 meter deep and running bank-full has an R of 0.33, an approximate V of 0.58, and a Q of 0.58. If the canal is widened so that Q is doubled to 1.16, A will become 1.71 and P will change to 3.71, with a resulting R of 0.46. Most important, V will increase to 0.68. With more rapidly flowing water come a myriad of problems for the irrigators, not the least of which is erosion.

One way of reducing the potential problems caused by an increased velocity would be to make the gradient less steep (International Land Development Consultants 1981:223). This can be done in one of two ways: by making the canal deeper in its upstream section than in its downstream reach, or by elevating the downstream end. The first of these options is rarely feasible. In most cases, canals that carry water out of ephemeral streams already have elevations at their heads that coincide with the bottoms of the natural channels. Similarly, canals leading away from perennial, or permanently flowing, streams, have their heads at approximately the same elevation as the water line. In both of these cases, there simply is not much room for a canal to be deepened.

Elevating the downstream end, including any new extension of the canal, is certainly more feasible and much more commonly done, than deepening the upstream end of a canal. In such cases, however, a sufficient amount of earth must be piled up to elevate the new segment; the amount of materials involved would be substantial. A 1-meter-by-1-meter canal, for example, might well need two or three times as much earth to support it above the normal ground surface as would be otherwise taken from an comparably sized excavated canal. Either of these approaches to reducing the velocity, however, would require an increase in the cross-sectional area, as Q is a function of both A and V.

A second way of decreasing the velocity of flow through a canal, thereby minimizing erosional problems, is to change the shape of its cross-sectional area. This is undoubtedly the most common method employed, although it is often used in conjunction with elevating downstream canal segments. For example, in order for a canal to have a Q of 1.16 with V of 0.58, it would have to have an A of 2.0 and a P of 6.0, so that R equals 0.33. The easiest way to achieve such proportions would be to fill in the existing canal and excavate a new one that measures 5.236 meters wide and .382 meters deep or one that is 5.2 meters wide across the top, 4.8 meters wide across the bottom, and 40 centimeters deep, with a step 20 centimeters high and 20 centimeters wide on each side. Either method obviously involves not only the reworking of a great deal of earth but refined hydraulic skills as well.

Extending canals or adding new ones to existing irrigation networks is, as should be apparent by now, a task more complex than many researchers have assumed. Not only do the new segments have to be added, but all the canals upstream of the new sections have to be enlarged and reshaped. The technology required for such changes often can be quite complex. In quantitative terms, doubling the size of an irrigation system involves much more than simply doubling the lengths of the canals. In terms of materials alone, the earth moved in expanding even a simple system would be at least three times as great. To illustrate, doubling the irrigated area by extending a canal measuring 1 meter wide, 1 meter deep, and 1,000 meters long would involve extending, perhaps by elevating, a new 1,000-meter-long segment of the same width and depth, and nearly doubling as well as changing the shape of the cross-sectional area of the original canal. Whereas the first canal required excavation of 1,000 cubic meters of earth in order to irrigate a given area, adding the new segment and enlarging the existing canal would require excavation of at least 2,000 additional cubic meters. All together, irrigating the initial area would involve the excavation of 1,000 cubic meters of earth for a canal, but irrigating twice the area would involve excavation of no less than a total of 3,000 cubic meters (see also Carruthers and Clark 1981 : 131–134).

If a small canal enlargement such as this involves a trebling of the amount of earth moved, a great deal more would be required for the expansion of a large canal irrigation complex, perhaps as much as eight to ten times the earth! If it could be quantified, the skill of the builders or their engineering expertise probably also would need to be more than twice as great. Given the magnitude of change in terms of the amount of materials involved and knowledge needed by the farmer-builders, it is likely that requisite changes in the social and political institutions were much greater than most researchers have considered.

Reasonably small systems can be rebuilt in brief periods of time, such as between growing seasons, with minimal amounts of both engineering expertise and labor. Small groups of even inexperienced irrigators working on a trial-and-error basis could make the necessary changes without losing any valuable time or land during that part of the year in which crops should be in the fields. Such small-scale systems worked so well and were so adaptable to modification as conditions and needs changed that they not only supported a large population through the Classic Period but were used throughout Mexico and functioned for centuries. Only when the population became extremely large and dense were demands for food so great that modifications could not be undertaken by small groups.

In her review of ideas concerning irrigation and cultural development several years ago, Barbara J. Price (1971:41) observed "that a kind of variable *critical-mass phenomenon* is evidently involved" (emphasis added). The tentative or cautious nature of her statement is undoubtedly a function of the relatively little systematic research conducted up until that time. Much additional work has, of course, been done since that statement was written. This synthesis of both the older and the recent research findings confirms what Price originally suspected. By late prehistoric times, demands for food in Mexico not only had reached a critical point but were so great that, by themselves, farmers could no longer make the necessary modifications to their field systems during the off-season. Improvements and enlargements were of such a scale that they would have required agriculturalists to spend part of the time they normally would have spent in their fields tending their crops during the growing season working on expanding canal networks. Taking farmers out of their fields for such activities would have the same effect as taking part of their fields out of production, surely a counterproductive action in the worst of times. Not only could farmers not afford to make such reallocations in their time and energy, but the necessary enlargements were of such a magnitude that they could not be carried out using trial-and-error construction techniques. A staff of learned professional civil engineers and a large labor force were needed.

7. Accomplishments and Contributions

The relationship between people and the biophysical environment in which they live has been the topic of numerous studies conducted during the past few decades. In such studies, the role of technology has been considered, but usually only implicitly. Concern for addressing it explicitly and systematically is, however, increasing. Indeed, a great deal of interest has been shown lately in understanding the impacts of technologically advanced cultures on the lands and people of technologically less developed ones. With the upcoming quincentennial of the discovery of the New World by people from southern Europe, it is particularly important to understand the nature of canal irrigation technology in Mexico prior to the subsequent modifications made by the Spaniards in the sixteenth century. It is also essential that the technological contributions to recent and present-day hydraulic agriculture made by the indigenous people be recognized.

What They Did and Did Not Know

By the end of the fifteenth century A.D. some rather elaborate canal irrigation systems had been developed in Mexico. For all the accomplishments, however, there were some technologies that were not discovered, or at least not used, presumably because they were not needed, or mastered. A clearer understanding of both what was actually achieved and what remained unutilized emerges from a review of technological developments that occurred in select elements of canal systems or some of the principal irrigation-related features themselves. Especially important are changes that took place in the channelization of streams and the construction of dams, canals, and other water control structures.

Channels

Developments made in channelization technology are as ancient as irrigation itself, and perhaps more so. If canal irrigation evolved out of floodwater farming, as one scheme suggests, then the very earliest canals probably were nothing more than modified gullies or small natural channels. If it developed out of drainage, as the other scheme holds, then, again, a situation existed in which natural watercourses were the precursors of canals. Regardless of the suspected origins, the channelization of stream courses is not younger than irrigation. It is either older than, or at least contemporaneous with, the earliest known irrigation. Indeed, the earliest confirmed and undisputed canals in Mexico are associated with channelization.

As will be recalled, the canals at Santa Clara Coatitlan carried water to fields directly from an artificially widened and deepened channel that collected ephemeral runoff from surrounding upland slopes and deposited it in Lake Texcoco. In all likelihood, the channelization at Santa Clara Coatitlan was not intended to facilitate the transport of water to the fields. Indeed, in a runoff situation such as this one, water would have been plentiful at certain times, and farmers would have planted after the ground had been amply wetted by sheetfloods. The earliest farmers at Santa Clara Coatitlan probably had more problems with damage to crops caused by runoff events later in the growing season than they had with water shortages. The channelization effort, therefore, might well have been one in which the natural stream bed was enlarged so as to better contain and direct water that otherwise flowed uncontrolled over a wide area. That the bottoms of the canals were higher than the thalweg, or the bottom, of the channelized stream suggests that water ran through the former much less frequently than it did through the latter. In effect, the channel fed the canals only some of the times that it carried water. Most of the time, it functioned to keep excess water off the fields.

Channelized streams became more complex over time, and the functions they served became increasingly varied. The first evidence of this comes from features found at Amalucan. There, as discussed in Chapter 3, the canal network identified by its discoverer as being one of irrigation appears to have been more of a drained field complex than a canal irrigation system. In addition to helping keep surplus water from entering the fields, therefore, the enlarged and straightened channel also collected excess from the fields.

Although the two systems functioned quite differently, the initial channelization effort at Santa Clara Coatitlan, and the technologically somewhat more complex and later one at Amalucan both involved the simple enlargement of natural stream channels. A similar alteration happened still later on the Tlajinga Plain just south of Teotihuacan. There, however, an entirely new channel was also excavated.

The canal irrigation system at Tlajinga functioned more like the older one at Santa Clara Coatitlan than the more recent one at Amalucan. The Tlajinga system appears to be one in which farmers began by practicing runoff agriculture. Later, they modified the ephemeral stream channel in order to better control the flow of water on the fields. Unlike the stream at Santa Clara Coatitlan that was channelized to carry water around and away from the fields, the lower end of the steam bed at Tlajinga was converted into a true irrigation canal. This was accomplished only with the coeval excavation of an entirely new channel that carried the normal stream flow around the cultivated area to another channel that paralleled the field on the side opposite that of the former channel that had been converted into a canal. In effect, the most important aspect of this channelization effort is that the ancient people permanently altered the normal course of stream flow. In so doing they not only were able to divert a percentage of the stream's flow to irrigate their fields, as earlier irrigators had done, but here they redirected the entire stream.

Perhaps not surprisingly, it did not take long for this technology to be put to use in irrigating larger tracts. With the ability to relocate stream channels came the ability to irrigate larger tracts of land as well as areas that were marginal for agriculture. One locale where evidence of such channelization was found is not far from Tlajinga. That site is the Maravilla system just to the west of Teotihuacan. There, an entire stream channel was relocated, but in this case it was done in order to carry water to fields rather than divert it away from them.

At first glance, this channel relocation might seem more like a canal than a channel. Certainly, the transport of water through an artificially excavated conduit to fields fits the literal definition of canal irrigation (see Chapter 1). On closer inspection, however, there is a subtle difference that is all-important. Canals involve the diversion and transport of only a small portion of the stream flow. Channels, in contrast, carry the preponderance, if not all, of a stream's normal discharge. While the Maravilla system does have a canal, it also involved an entirely relocated stream channel.

In addition to transporting water to fields, the channelization evident at Maravilla is also important on another front. This effort not only involved the excavation of a new stream channel but also involved the construction of sizable earthworks in order to insure that the stream continued to follow its new course and did not readjust itself back into the former channel. The builders of the Maravilla system obviously applied tried-and-proven dike and dam technology to the building of channel relocation structures.

Without doubt the greatest channel relocation technology evident anywhere in prehistoric Mexico was the channelization that was done near Cuautitlan. In one respect this project was similar to that of Maravilla: it involved diverting the entire stream flow to rather than away from a place. In all other respects, however, the system was different. The Cuautitlan channelization extended for several kilometers, not meters. It involved, for the first time, a perennial river, not an ephemeral stream. Because of its size and the nature of the diverted stream, it required the construction of some truly monumental earthworks. Lastly, and perhaps most important, it carried water from one valley to another. Much could undoubtedly be said about the significance of the technology developed and incorporated here had the documentary and archaeological sources revealed more. From what little is known, however, one is conservative in concluding that, by anyone's standards, this channelization effort was one of the more herculean feats of hydraulic engineering anywhere in Mexico. The expertise evident here rivals that of the best civil engineers working today.

The developmental sequence of channelization reconstructed and outlined here shows a logical progression in terms of both technological complexity and the skills of the builders. Although there are only a few cases in which channelization has been confirmed archaeologically, the data clearly indicate how the technology was developed. They also demonstrate that ancient Mexicans were proficient in controlling the flow of streams. In not one case is there evidence of channelization not working successfully for a long period of time. The closest thing to a failure was the first known channelization effort at Santa Clara Coatitlan. Sedimentation problems evident there, however, might well be chalked up to the infancy of the technology. Even if this is the case, the system functioned for nearly two centuries.

Dams

Writing as recently as the past decade, one authority on the history of hydraulic technology, Norman Smith (1971 : 144), noted that "There is no lack of evidence that the Spaniards brought dam-building and its associated technologies to the Americas." This statement might well sum up the beliefs of many people. Unfortunately, however, it appears to reflect a Eurocentric bias on the part of the writer. There is little reason to suspect Smith of attempting to provide something other than the truth as he interpreted it. It is possible that he did not know of the prehistoric dams in Mexico. The body of literature on such features was certainly small at the time, and details were hardly convincing (e.g., Armillas, Palerm, and Wolf 1956). If Smith did know of them, he probably discounted the importance of such features. To be sure, dams known to have been used for irrigation purposes in Mexico during ancient times were neither as large and as technologically complex nor as economically important as many of those used in the Old World (e.g., García-Diego 1977). Dams in Mexico tended to be small and comparatively simple in terms of engineering and construction, and they facilitated watering lands that sustained only a small percentage of the total population. Nevertheless, evidence of nearly every type of dam known from anywhere in the world does exist in prehistoric Mexico, and these can be dated rather accurately. In some cases, the dates at which various technologies were initially developed cannot be confirmed. Inferences drawn from well-documented associated features, however, often corroborate the ages of specific items and allow for them to be assessed in a developmental context.

Dams that stored water from ephemeral sources and those that diverted it from ephemeral as well as perennial streams both existed in Mexico prehistorically. Given the two possible, but thus far unsubstantiated, scenarios for the origin of canal irrigation, it is impossible to say which preceded which or if they were developed independently of each other. If the dam at Teopantecuanitlan is eventually confirmed as having served an irrigation function, and no earlier dams are found, then it can be safely concluded that storage dams were developed first. The simplicity of brush and stake weirs, combined with their well-demonstrated sequence, can, however, be used in arguing a greater antiquity for diversion structures. Certainly, more data need to be collected, and much more work needs to be done in order to unravel the series of events and their connections. For now, however, at least a few conclusions can be drawn about dams and the development of their technology.

Regardless of the various possible twists in arguments about origins, the earliest dams involved the use of materials close at hand and with minimal, if any, modification. The dam at Teopantecuanitlan involved nothing more than the stacking of rough, uncut, and unfaced rocks across an ephemeral channel. The Purron Dam was an earthen structure that presumably required nothing more than piling and perhaps compacting. Both of these structures were approximately twice as wide at their bases as they were high, and situated perpendicularly to the stream channels. They are classified in technological terms as "gravity dams" (Arthur 1965:64; N. Smith 1975:228) because they depended on their own weight for support against the weight of the water they impounded.

The dam at Monte Albán Xoxocotlan is clearly of another generation, as it shows marked technological advancements over its two known earlier counterparts. This dam was made largely of piled, unmodified rocks. However, the uppermost portion was made of cut and fitted blocks, and the entire structure was plastered over, thereby providing a certain degree of watertight integrity. The V-shape of this dam with the apex pointing upstream indicates that its builders succeeded in freeing the dam from a dependency on volume and bulk in order to support the force of impounded water. Technologically, this feature can be classified as an "arch dam" (Arthur 1965:66; N. Smith 1975:227). The arched plan in conjunction with some cut and fitted rocks, and the plastering, all resulted in a dam that was physically smaller than it would have been otherwise. Accordingly, this dam was nearly as high as it was wide. The rocks saved as a result of this advanced technology were presumably used in the construction of the monumental buildings at the Monte Albán site.

Dam technology seems to have reached a zenith ca. 550 B.C. Indeed, it is not until A.D. 400 that evidence of dams appears once again. Such an extended period suggests not only that this type of structure fell out of use, but also that the technology involved was lost. The nature of the dams when they did reappear tends to confirm this conclusion.

The next dams to be built in Mexico were those across the Río Xiquila and in Tecorral Canyon. Evidence, albeit tentative because the actual prehistoric structure has been long destroyed, indicates that the former consisted of little more than rocks piled into this perennial stream. The latter was somewhat more complex, involving an earth and rock-rubble core, but it was faced with dry-laid unworked cobbles and boulders. The use of earth in the construction of this dam is undoubtedly a function of its being in an ephemeral stream channel. The rock riprap on the face was, apparently, to stabilize the structure and insure that the force of flash floods in such environs did not destroy the dam. Earth could not have been used in the Xiquila dam, as that one was built across a perennial stream, and a rather large one, at that.

In both the Xiquila and the Tecorral cases, sheer bulk was the single factor responsible for anchoring the dams in place. However, although they can be considered "gravity dams," they were not built perpendicular to the stream flow as is normally the case with dams so classified. Indeed, both of these features were constructed diagonally across their respective channels, with the downstream ends located near the mouths of the main canals. Unlike the earlier dams at Monte Albán and in the Tehuacan Valley, the Xiquila dam was clearly a diversion dam, whereas the one in Tecorral Canyon not only diverted water but also stored it. Both appear to have been developed as much out of earlier rudimentary forms of ephemeral stream diversion technology as out of water storage.

Exactly what kinds of improvements were made in dam technology during the following centuries and prior to the arrival of the Spaniards is unclear. It is known that rock diversion dams probably no larger or better built than that found in the Río Xiquila, and certainly smaller and not as complex as those used today, were built across the Río Tula and some of its perennial tributaries ca. A.D. 1000.

Shortcomings and lack of achievement are certainly evident in the dams used prehistorically in Mexico. Many traits that are commonly associated with dams throughout the world did not appear in the region until after the Spaniards appeared (Table 7.1). For example, although the ancient Mexicans were accomplished stonemasons who cut, fitted, and mortared stones in constructing buildings, they never applied that technology to dams used for irrigation. One report (Abascal and García Cook 1975 : 202) does, of course, argue the contrary. However, as discussed in Chapter 1, there are severe problems with the way those features were analyzed and dated. In all likelihood those dams either were Spanish additions to prehistoric canals or were not described accurately.

At least three other aspects of common dam construction were not developed prehistorically in Mexico. First, "core-walls" (N. Smith 1975 : 228, 229), or watertight barriers, were not integrated into either earthen or rock dams. The closest that ancient Mexicans came to doing this was the building of an earth and rubble core dam in Tecorral Canyon. The plaster covering on the dam at Monte Albán Xoxocotlan also probably helped in reducing the permeability of that dam. Second, "buttress dams" (Arthur 1965 : 67; N. Smith 1975 : 227), those that involved buttresses on the downstream face to help support the weight of impounded water, are not evident anywhere in Mexico prior to the sixteenth century. Buttresses were, of course, used to support the canal walls on the slopes of Cerro Purificación. There should be little doubt, therefore, that they would have been used on dams in late pre-Hispanic times had large masonry dams existed then. Finally, storage dams were not built across perennial

Table 7.1. *Technology not developed as part of canal irrigation in prehistoric Mexico*

Dams	Other Water Control Devices
Cut, fit, and mortared stone construction	Board-type head and sluice gates
Core-walls	Arched aqueducts
Buttress dams	Flow control boxes
Storage dams on perennial streams	Drop structures
	Chutes
	Pipes
	Inverted siphons

streams in Mexico until after the Spaniards arrived. This condition is probably a function of demand, the lack of which is also related to the absence of core-walls and buttresses. Put quite simply, ancient Mexicans had little need or desire to dam up permanently flowing streams for irrigation purposes. From all accounts, they appear to have developed ways of providing an abundance of food without storing large amounts of water. Had they ever needed to do so, they surely would have, as the required elements of the needed technology had been developed and were available for incorporation.

The technology involved in building dams certainly has a checkered past in the prehistory of Mexico. Many significant things were achieved only to be lost and then redeveloped. To be sure, some dam-related technologies were never developed. However, ancient Mexicans did develop and use dams. These accomplishments should not be discounted and they cannot be overlooked.

Canals and Other Water Control Devices

The distinguishing features that characterize all canal irrigation systems are, as the term clearly indicates, the canals themselves. Although they all serve the same general purpose—to transport water under the force of gravity—there are notable differences in the specific functions of various canals. For example, main canals carry water directly away from the source to branch canals that, in turn, feed field canals that distribute water evenly over the planting surfaces. In addition to function, canals differ considerably in length, width, depth, cross-sectional shape, wetted perimeter, gradient, roughness, and sinuosity. All of these factors affect not only the velocity at which irrigation water flows through the system, but also the rate of discharge and, ultimately, the amount of land that can be irrigated.

As will be recalled from the previous chapter, the hydraulics of any canal system can be calculated mathematically if these variables are known. Armed with the results of such computations, one could evaluate the relative efficacy of each system by comparing it to others. Assuming, as per the premises outlined in Chapter 1, that higher degrees of efficacy were sought by irrigators, a comparison of this nature would be ideal for assessing the development of canal technology over time.

Unfortunately, data on prehistoric irrigation canals in Mexico are not sufficiently complete for an undertaking of this type. The greatest single deficiency is with canal gradients. Researchers responsible for the collection of detailed information on ancient canals have done an admirable job of reporting the cross-sectional measurements of canals, undoubtedly because these are the very types of data that are easily extracted through archaeological excavations. They have, however, been remiss in not recovering data on canal gradients. Canal lengths have, of course, been reported in almost every case, but this information is of limited value for understanding the hydraulic efficacy of a system if elevations at noted points along the canal are not given. For each of the prehistoric irrigation canals known in Mexico and analyzed here, an attempt was made to determine elevations from topographic maps. This procedure proved fruitless unfortunately, because, at 1:50,000 scale, the best available maps have contour intervals of 20 meters, making the margin for error far too great for gradients to be determined.

Although the data are not suitable for a detailed analysis of development based on hydraulic efficacy, there is at least one conclusion that can be made about ancient hydraulic technology in Mexico: it was learned early on that the cross-sectional shape of canals played as great a part in the functioning of irrigation systems as did size. The four earliest canals for which cross section details are known, the ones at Teopantecuanitlan, Santa Clara Coatitlan, Tlaxcala, and Monte Albán Xoxocotlan, all had different shapes—rectangular, trapezoidal, U-, and stepped, respectively. It has been noted already that this variety indicates that much experimentation was under way. Significant now, however, is that no single shape was recognized by prehistoric irrigators as the best or most appropriate and, hence, used exclusively from the time it was developed. On the contrary, all of these shapes, as well as at least one new one, were used in later prehistoric times. For example, rectangular canals were used again in the Xiquila Valley, trapezoidal ones were used again at Amalucan, and U-shaped ones were used later on the Tlajinga Plain. V-shaped canals were first used at Otumba and then again at Tula. Clearly, the lack of any one consistently used cross-sectional shape suggests that canal builders did whatever was necessary in terms of adjusting hydraulic variables in order to get their systems to function properly.

Size is conspicuously important for maintaining a given canal discharge. Although shape might not be as apparent, it was a variable of known importance to ancient irrigators. In many cases, such as short, narrow, steep-sided valleys, irrigators were restricted to building canals of given lengths and, hence, gradients. Although they could easily build a canal with any needed cross-sectional area, maintaining the appropriate velocity and, hence, discharge was problematical. Especially critical in this regard is the hydraulic radius, which, holding the area constant, is a function of the wetted perimeter. This last variable is, of course, a function of cross-sectional shape. For example, a canal that is 1 meter deep and 1 meter wide has a cross-sectional area of 1 square meter and a wetted perimeter of 3 meters. A canal that is 25 centimeters deep and 4 meters wide also has an area of 1 square meter, but it had a wetted perimeter of 4.5 meters. This latter canal obviously has more friction-creating surface and, hence, under otherwise equal conditions, will have a slower velocity and a lesser discharge than the former canal.

Mexican irrigators mastered the rudiments of canal hydrology early on in prehistoric times. There was, therefore, little they could do later to improve the mechanics of getting water to flow through their canal networks efficiently. They did not, however, stop making technological developments. Indeed, as different types of water sources were being tapped, as different environments were being cultivated, and as irrigation systems became more complex with the degree of branching becoming greater, canal builders had to devise a number of features to control the flow of water. Not surprisingly, these developments grew out of each other.

The earliest known water control devices are the earthen obstructions found in the field canals at Santa Clara Coatitlan. Features such as these certainly took little skill in constructing. They do, however, indicate quite clearly that irrigators even at this early date were dealing with problems of getting water out of canals and manipulating the flow of it across fields. Although piling earth in small canals is an elementary achievement, it was probably preceded by even more simplistic brush devices such as those discussed in conjunction with canals originating from the use of floodwaters.

Blocking field canals with earth, like most other technological accomplishments, not only had some less elaborate predecessor but also led to other more elaborate achievements. Specifically, earthen constructions built across field canals in order to distribute water over planting surfaces at Santa Clara Coatitlan were the antecedents for somewhat larger obstructions across larger canals. Indeed, that appears to have been exactly what was done a short time later at Monte Albán Xoxocotlan. There, water flowing through the system had no known way of getting out of the main canal unless some type of temporary obstacles were built at critical points. In effect, such earthen fill can be considered an incipient form of sluice gate. What keeps them from being true sluice gates is that they were not used to regulate the flow of water from one canal to another, nor did they funnel water out of canals through openings or, quite literally, "gates."

The first true sluice gates were used at Hierve el Agua. There, because perennial water and branch canals were used together for the first time, it was imperative that obstructions be placed across the mouths of canals at critical places and times in order to divert the flow of water to wherever it was needed, when it was needed. Conversely, such features were also used to keep water out of areas where it was not needed.

Hierve el Agua appears to have been a pivotal point for developments in sluice gate technology. Although they were small, their degree of branching minimal, and the flow of water through them slight, these canals provided irrigators the opportunity to perfect the techniques for manipulating permanently flowing water. These farmers were free from having to deal with the flash floods often associated with ephemeral streams, and they did not have to concern themselves with diverting irrigation water out of large, rapidly flowing perennial streams or rivers. The technology they developed, however, was certainly essential for later irrigators who had to deal with the latter problem. In effect, just as the irrigators at Santa Clara Coatitlan had developed devices that helped their later counterparts at Monte Albán Xoxocotlan to get water out of canals, the irrigators at Hierve el Agua developed the technology for diverting permanently flowing water.

The sluice gate technology developed at Hierve el Agua, from the evidence that now exists, appears to have evolved into head gate technology for the first time at the Loma de la Coyotera site. There, irrigators were faced with a problem exactly opposite that faced by earlier irrigators who diverted water out of ephemeral streams. Whereas their earlier counterparts worked to get water to flow from the channels into the canals, irrigators at Loma de la Coyotera worked most of the time to keep water in the channel and out of the canal. It was only when it was needed for irrigation purposes that the canal was opened and water allowed to flow to the fields.

Controlling large volumes of flowing water required that head gates such as those used at Loma de la Coyotera be substantial features. No archaeological evidence exists to indicate what these gates looked like but, if ethnographic analogs are at all acceptable, they involved little more than a suitable amount of earth back-filled and compacted. There is nothing to suggest that any type of formal gate structures, such as pairs of vertical posts embedded on both sides of the canal, between which planks could be stacked, were used in Mexico prehistorically. The closest things to such devices are the notches found in the sides of the travertine-encrusted canals on the Llano de la Taza. Exactly when these gates were constructed is not known because of the long period of time over which the system was used and expanded, and because techniques for dating travertine have not been perfected. Given the size of the system and its complexity, as measured by the high degree of branching, it can be safely concluded, however, that notched canals are a relatively late accomplishment in the development of sluice gate technology. If permanent wooden plank-type gates were ever to be used prehistorically in Mexico, it is most likely they would have developed out of the notching noted in the Llano de la Taza canals.

Just as the Hierve el Agua site played a key role in the development of sluice gate technology, the Loma de La Coyotera site was important not only because of the developments noted there in head gates, but also because it is the earliest site from which aqueducts are known in Mexico. Without going into as much detail as was done with channelization and head and sluice gates, suffice it to say that aqueduct technology went through at least three distinct phases of development. The earliest ones were the short, low earthen structures built in order to carry water over the upstream end of small ephemeral channels. In effect, these aqueducts did little more than shorten the total length of canals had the latter followed the natural contours around the basins of the drainages they crossed. These structures appear to have led to the development of technology used in both building up canals along the sides of steep slopes and trellis-type aqueducts in the Xiquila Valley. Just as the Loma de la Coyotera aqueducts filled the upper ends of small drainages, the Xiquila canals involved filling some gullies in the hillsides. Similarly, just as the Loma de la Coyotera aqueducts crossed small drainages, the trellis-type ones used in the Xiquila Valley crossed much larger ones.

Development of trellis-type aqueducts led to the later development of combination trellis-type and low masonry aqueducts crossing Lake Texcoco between Chapultepec and Tenochtitlan. From there, tall, steep-sided, masonry and stuccoed aqueducts were developed at Cerros Purificación and Tetzcotzingo. Had Spaniards not shown up in the sixteenth century, the Aztecs probably would have gone on to develop arched aqueducts not unlike those that characterize the landscape of much of the region that made up the core of colonial New Spain. Indeed, they had, as part of the aqueducts built in the Basin of Mexico during the fifteenth century, developed something that was later a hallmark characteristic of large Spanish irrigation features—lined canals.

Although the depths and the widths of the canals associated with these aqueducts were small by any standards, they were the first to show signs of having been lined. Presumably this was done to reduce (1) permeability and, hence, loss of scarce and therefore valuable water to seepage, (2) roughness, thus speeding up the velocity of flow in order to minimize evaporation in transit, (3) erosion of canal sides and downcutting, or (4) maintenance associated with cleaning and removing sediment and other debris. Whatever the specific reasons might have been for lining canals, the act of doing so was a technological achievement developed out of mortaring and stuccoing, and it is known only from this place and time.

There are no existing data to suggest what, if any, type of sluice gates were used as part of the lined canals in the Texcoco area. The only information available notes that water was taken out of the canals through cut rock troughs. No evidence of post holes or slots for boards or planks has been reported. Presumably, therefore, sluice gates of such a nature were not involved. In all likelihood, given the materials used in construction and the known masonry skills of builders in the area, rocks cut to fit the inside of the canals were used. Such items could have been inserted and removed quite easily as needed, and would have worked most effectively.

In sum, there were very few things pertaining to the technology of irrigation canals that were not developed in Mexico prior to the arrival of Spaniards in the sixteenth century. Granted, a few of these accomplishments were very late and not widely known and used. They did, however, exist. About the only things that were not developed prehistorically were board-type sluice gates, arched aqueducts, and an assortment of masonry devices such as flow control boxes, drop structures, chutes, pipes, and inverted siphons (see, e.g., Zimmerman 1966:118–123, 260–262, 297–305, 310–313, 341– 344), which are used to regulate the flow of water through various parts of canal irrigation networks (Table 7.1). The inverted siphon (see, e.g., N. A. F. Smith 1976) is found throughout Mexico today, and is used most frequently when trellis-type aqueducts would otherwise have to be used. These features, and probably all other canal irrigation-related structures found in Mexico today and not discussed herein, were introduced in historic or modern times.

Post-contact Renovations

It has been reported in at least two separate places (Olin 1913:9, Kinney 1981:87–88) that the army of Hernán Cortés was impeded in its conquest of Mexico between 1519 and 1521 by the numerous irrigation canals it had to cross. The exact origin of these reports remains unknown. Cortés' own correspondence (Morris 1928:59, 151), his secretary's chronicles (López de Gómara 1943:183, 230), and the account of events by Bernal Díaz (1963:291, 347, 349) mention canals only a few times. Although the reports of canals being repeated obstacles to travel remain unsubstantiated, Spanish documentation from later in the sixteenth century is of sufficient quantity and quality to suggest that such was at least possible. Indeed, there exists abundant evidence demonstrating that canals were numerous and that their use in agriculture during pre-contact times was widespread (Wolf 1959:76), including even areas where archaeological evidence of canals has yet to be discovered.

The credit for first determining the regional distribution of irrigation in Mexico at the time of contact, and, hence, demonstrating its importance prehistorically, goes to Angel Palerm. In what must have been a time-consuming and labor-intensive activity, he gleaned from an unknown number of sixteenth-century documents in the Archivo General de la Nación de México references to no less than 376 locations where irrigation was reported as having been practiced by indigenous people when the first Spaniards arrived (Palerm 1954; 1961a). Although most of the places are on the Mesa Central, irrigation was reported as far away as present-day Nayarit state in the northwest, Chiapas in the south, and Veracruz on the Gulf Coast (Fig. 7.1). More recent studies have extended this range to include an outlier far to the northwest in Sonora (Doolittle 1984a; 1988).

Many of the sites identified by Palerm probably did not involve

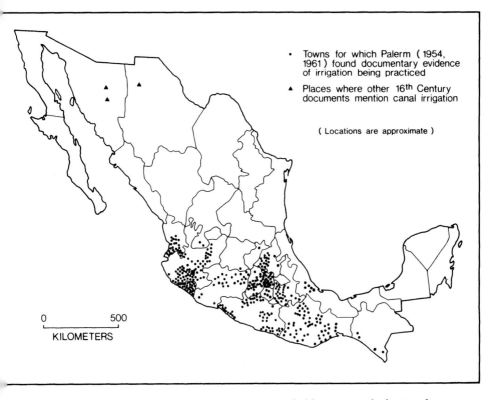

Figure 7.1. Locations of canal irrigation as recorded by Spaniards during the sixteenth century A.D.

canal irrigation per se, but other forms of agricultural water control instead. Indeed, seven locales were specifically identified as having chinampas. Others might well have been floodwater farming sites, and some might have relied on manual irrigation with water taken from shallow wells. Regardless of such exceptions, irrigation was clearly widespread, and canals were used extensively for transporting water to crops at the time of European contact.

This practice was so obviously important to sustaining large numbers of people that the Spaniards moved preferentially, and quite quickly, into areas irrigated by indigenous people. Irrigation was also so widespread that its existence facilitated the rapid expansion of the Europeans and their dominance of the region. Although the Spaniards eventually moved into previously unirrigated or only marginally irrigated areas and constructed entirely new canal networks as a result, their principal approach to agricultural settlement involved taking over existing canals built by people in pre-Hispanic times. As is so often the case, this situation has been described no more eloquently than by Eric Wolf:

> . . . the Spaniards also laid hands on the scarcest and most strategic resource in Middle American ecology: water. They needed water to irrigate newly planted fields . . . Sons of a dry land themselves, they were master builders of aqueducts and wells; but all too often, they appropriated the canals of the native population . . . (Wolf 1959:199)

Wolf's statement was made in reference mainly to the situation of the Mesa Central and points to the south and west. It has relevance to the whole of Mexico, however, including the north, where Spanish settlement occurred later on. Examples of Spaniards taking over canals built in pre-contact times have been reported from the present-day states of Zacatecas, Sonora, and Chihuahua. In some cases (see, e.g., Sayles 1936:38–39, 41, 88; Trombold 1985:247), these involved canals that were quite ancient and had been abandoned long before the Spaniards arrived. Apparently Spanish settlers found the canals to be in relatively good shape, hydrologically sound, and compatible with their own technology. Accordingly, they seem to have done little more than clean them out, make some basic repairs, and press them into service. In other cases (see, e.g., Hinton 1983:317), Europeans moved in and took control of canals that indigenous people not only had built, but were still using.

The moral implications of this latter circumstance are profound, but beyond the scope of this study. The fact that Spaniards not only took over ancient systems, but saw indigenous people actually using them is, however, significant in terms of understanding the development of hydraulic technology in the New World.

As both Smith's earlier quote about dams and Wolf's above quote about aqueducts clearly indicate, Spaniards have been long perceived as being superior to indigenous farmers in terms of agricultural engineering (e.g., Trautmann 1986:246). This attribute probably stems from their accomplishments in building hydraulic features in Mexico and the existence of similar structures in Spain. In fact, however, some, perhaps most, of the Spanish masons responsible for building such things as arched aqueducts in Mexico had no previous experience (Kubler 1944:9, 16). This probably holds especially true for the missionaries who did such work. Although there currently exist no studies to demonstrate this, it may well be that the aqueduct builders in Mexico simply tried to mimic those they were familiar with in Spain. There is only one documented case, that involving the mission at Huejotzingo in present-day Puebla state, in which the builder of the aqueduct was an experienced craftsman who came from Spain in the middle of the sixteenth century (Prem 1975 : 33).

Additional evidence for the less than advanced state of knowledge and expertise about irrigation held by the Spaniards comes from a recent study conducted in the Bajío and involving mainly the seventeenth century. There, Michael E. Murphy (1986 : 175) found that in terms of technology, Mexico was only partially connected with the European centers of innovation. This was especially true in the more remote parts of the region, where travel and, hence, communication were exceedingly slow. Murphy found, as did George Kubler, that only rarely did the persons responsible for initiating and overseeing the construction of irrigation systems have any formal training. For the most part, they simply did what they thought would work (see also Kubler 1944 : 17). They also learned a great deal from indigenous people who themselves were experienced irrigators.

It is well recorded that the native labor used in building aqueducts for the Spaniards in the sixteenth century came from Tenochtitlan and Texcoco (Kubler 1944 : 18). Indeed, Fray Tembleque, who oversaw construction of a most impressive aqueduct (Fig. 7.2) in the 1540s, purportedly with no previous experience (Romero de Terreros 1949 : 40), utilized workers from the latter area. That these workers contributed technological insight into the building of such features is evident in the details of construction. For example, small rock fragments were typically wedged into the chinks between the larger irregularly shaped stones of so-called Spanish-built aqueducts in Mexico and then mortared over (Fig 7.2). This particular practice was used in pre-Hispanic aqueduct construction both at Chapultepec and in the Texcoco area. It was used again, at least in part, in the historic-era reconstruction of the Chapultepec aqueduct (Baxter 1934 : 128; Chapultepec: Historia y presencia 1988 : 55) and in construction of the eighteenth-century aqueduct at Queretaro (Fig. 7.3) (Septién y Septién 1988 : 15–17). It was not, as far as available evidence indicates, used anywhere in Spain, where aqueducts were built of regularly faced and tightly fitted stone blocks (e.g., Fernández Ordóñez et al. 1986). The monumental hydraulic features and associated canal irrigation systems that characterize the landscape of much of central Mexico are, therefore, as much a function of the expertise, knowledge, and work of indigenous people as they are of the Spaniards who until now have been given total credit for

Figure 7.2. View of an aqueduct built in the latter half of the sixteenth century under Spanish supervision near the present-day town of Zempoala, Hidalgo (not to be confused with the pre-Hispanic site of the same name in Veracruz state). Note the plaster covering rock fragments used to fill the chinks between the irregularly shaped rocks.

Figure 7.3. Close-up view of an aqueduct built to carry water to a Franciscan college in the city of Queretaro during the late 1700s. Note the rock fragments wedged into the roughly cut and poorly fitted rocks.

their development. Whereas the architecture of such aqueducts is undoubtedly Spanish, the engineering is clearly indigenous. The Spaniards' first choice of agricultural land was that already irrigated by the indigenous people. When those lands were completely taken over, the Spaniards expanded their operations into previously non-irrigated areas and relied heavily on both the labor and the skills of native Mexicans to build and operate canal irrigation systems.

The same things that were just said of the relationship between pre-Hispanic and Spanish canal irrigation can be said of many present-day and ancient systems. From the time of the very first systematic studies of prehistoric irrigation in Mexico (e.g., Wolf and Palerm 1955 : 269), it was clear that canals carrying water to crops today did so long ago as well. This situation holds especially true for the Basin of Mexico. There, canals have long histories of continuous use, starting in prehistoric times and extending through the colonial period right up to the present (Sanders 1976 : 116; Sanders, Parsons, and Santley 1979 : 263). Canal systems with such prolonged sequences have been identified in the Teotihuacan Valley (Armillas 1961 : 266), the Texcoco area (Parsons 1971 : 146), and near Cuautitlan (Rojas Rabiela, Strauss K., and Lameiras 1974 : 8).

In some cases, the exact history of irrigation technology is unclear, but canals built in relatively recent times are morphologically and functionally similar to ones built prehistorically. Furthermore, they are often located in areas known to have had large prehistoric populations. Foremost among these are the canal systems known from Tecomatepec in the state of Mexico (Palerm 1955 : 30), the Cerro Gordo area (Sanders, Parsons, and Santley 1979 : 266), and Tula (Mastache de Escobar 1976 : 64). There are other cases in which long-abandoned prehistoric canal systems have been rebuilt recently (Armillas, Palerm, and Wolf 1956 : 396), or in which new canals have been built on top of (C. E. Smith 1965 : 73) or adjacent to (Woodbury and Neely 1972 : 133) ancient ones. Lastly, there are cases in which canal irrigation networks built in prehistoric times are known to have been improved with modern technology. Perhaps the best case of such an occurrence involves the Tehuacan Valley, where some prehistoric canals have been lined with concrete (C. E. Smith 1965 : 68).

In conclusion, canal irrigation technology in Mexico has a long and illustrious past. Its sequence of development began perhaps earlier than that of agriculture itself. It involved a great deal of innovation, and much exchange of information by numerous people. The course of development was at times a circuitous one. Not only were there as many failures as successes, but developments often involved distant people in disjunct areas in unforeseeable ways. The result of all this, however, was the creation of a means by which a large population could be supported, free of the vagaries of nature, and without the need for intervention by deities. The success of prehistoric irrigators is perhaps best reflected in their contributions to economic development (albeit that of others) in colonial times and to the support of Mexico today.

Bibliography

Abascal, Rafael, and Angel García Cook
1975 Sistemas de cultivo, riego y control de agua en al área de Tlaxcala. In *XIII Mesa Redonda*, Sociedad Mexicana de Antropología. *Arqueología* 1:199–210.

Ackerly, Neal
1988 Prehistoric Agricultural Activities on the Lehi-Mesa Terrace. Hohokam Irrigation Cycles: A.D. 700–1100. Manuscript on file, Northland Research Inc., Flagstaff, Ariz.

Adams, I. H.
1976 *Agrarian Landscape Terms: A Glossary for Historical Geography.* London: Institute of British Geographers, Special Publication 9.

Adams, Richard E. W.
1977 *Prehistoric Mesoamerica.* Boston and Toronto: Little, Brown and Company.

Adams, Robert McC.
1968 Early Civilizations, Subsistence, and Environment. In *Man in Adaptation: The Biosocial Background*, edited by Yehudi A. Cohen, pp. 363–377. Chicago: Aldine.

Anderson, Edgar
1954 *Plants, Man, and Life.* Berkeley and Los Angeles: University of California Press.

Armillas, Pedro
1949 Notas sobre sistemas de cultivo en Mesoamérica: Cultivos de riego y humedad en la cuenca del Río de las Balsas. *Anales del Instituto Nacional de Antropología e Historia* 3:85–113.

1961 Land Use in Pre-Columbian America. In *A History of Land Use in Arid Regions*, edited by L. Dudley Stamp, pp. 255–276. Paris: UNESCO, Arid Zone Research 7.

1971 Gardens in Swamps. *Science* 174:653–661.

Armillas, Pedro, Angel Palerm, and Eric R. Wolf
1956 A Small Irrigation System in the Valley of Teotihuacan. *American Antiquity* 21:396–399.

Arthur, H. G.
1965   Selection of Dam Types. In *Design of Small Dams: A Water Resources Technical Publication of the U.S. Department of the Interior*, pp. 63–67. Washington, D.C.

B. de Lameiras, Brigitte
1986   El riego y el estado en el México prehispánico. In *La sociedad indígena en el centro y occidente de México*, edited by Pedro Carrasco, pp. 23–41. Zamora: El Colegio de Michoacán.

Bakewell, P. J.
1971   *Silver Mining and Society in Colonial Mexico, Zacatecas, 1546–1700.* Cambridge: Cambridge University Press.

Bandelier, A. F.
1892   *Final Report of Investigations among the Indians of the Southwestern United States, Carried on Mainly in the Years from 1880 to 1885, Part II.* Papers of the Archaeological Institute of America, American Series 4.

Barlett, Peggy F., ed.
1980   *Agricultural Decision Making: Anthropological Contributions to Rural Development.* New York: Academic Press.

Baxter, Silvestre
1934   *La arquitectura hispano colonial en México.* Mexico City: Departamento de Bellas Artes.

Beals, Ralph L.
1932   The Comparative Ethnology of Northern Mexico before 1750. *Ibero-Americana* 2:93–225.
1945   *The Contemporary Culture of the Cahita Indians.* Washington, D.C.: Smithsonian Institution, Bureau of American Ethnology, Bulletin 142.

Blanton, Richard E.
1972   Prehispanic Adaptation in the Ixtapalapa Region, Mexico. *Science* 175:1317–1326.
1978   *Monte Albán: Settlement Patterns at the Ancient Zapotec Capital.* New York: Academic Press.

Blanton, Richard E., and Stephen A. Kowalewski
1976   The Valley of Oaxaca Settlement Pattern Project: A Progress Report. On file, Department of Anthropology, Hunter College, City University of New York.

Bolton, Herbert Eugene, ed.
1908   *Spanish Exploration in the Southwest, 1542–1706.* New York: Barnes and Noble.

Boserup, Easter
1965   *The Conditions of Agricultural Growth: The Economics of Agrarian Change under Population Pressure.* Chicago: Aldine Publishing Co.
1981   *Population and Technological Change: A Study of Long-Term Trends.* Chicago: University of Chicago Press.

Bowen, Thomas G.
1976   Esquema de la historia de la cultura trincheras. In *Sonora: Antropología del desierto; Primera Reunión de Antropología e Historia del Noroeste*, edited by Beatriz Braniff C. and Richard S. Felger, pp. 267–279. Centro Regional del Noroeste, Colección Científica 27. Hermosillo.
n.d.   A Survey and Re-evaluation of the Trincheras Culture, Sonora, Mexico. Manuscript on file, Arizona State Museum, Tucson.

Brand, Donald D.
1933   The Historical Geography of Northwestern Chihuahua. Ph.D. Dissertation, Department of Geography, University of California, Berkeley.
1937   *The Natural Landscape of Northwestern Chihuahua. University of New Mexico Bulletin* 316, Geological Series 5.
1939   Notes on the Geography and Archaeology of Zape, Durango. In *So Live the Works of Men: Seventieth Anniversary Volume Honoring Edgar Lee Hewett*, edited by Donald D. Brand and Fred E. Harvey, pp. 75–105. Albuquerque: University of New Mexico Press.
1943   The Chihuahua Culture Area. *New Mexican Anthropologist* 6–7:115–158.

Braniff, Beatriz
1974   Oscilación de la frontera septentrional mesoamericana. In *The Archaeology of West Mexico*, edited by Betty Bell, pp. 40–49. Ajijic, Jalisco: Sociedad de Estudios Avanzados del Occidente de México.

Braniff de Torres, Beatriz, and María Antonieta Cervantes
1966   Excavaciones en el antiguo acueducto de Chapultepec I. *Tlalocan* 5:161–168.
1967   Excavaciones en el antiguo acueducto de Chapultepec II. *Tlalocan* 5:265–266.

Bribiesca Castrejón, José Luis
1958   El agua potable en la República Mexicana: Los abastecimientos en la época prehispánica. *Ingeniera Hidráulica en México* 12.2:69–82.

Brown, R. B.
1985   A Synopsis of the Archaeology of the Central Portion of the Northern Frontier of Mesoamerica. In *The Archaeology of West and Northwest Mesoamerica*, edited by Michael S. Foster and Phil C. Weigard, pp. 219–275. Boulder and London: Westview Press.

Brundage, Burr Cartwright
1979   *The Fifth Sun: Aztec Gods, Aztec World*. Austin and London: University of Texas Press.

Brunet, Jean
1967   Geologic Studies. In *The Prehistory of the Tehuacan Valley*, vol. 1, *Environment and Subsistence*, edited by Douglas S.

Byers, pp. 66–90. Austin: University of Texas Press for the Robert S. Peabody Foundation.

Butzer, Karl W.

1982   *Archaeology as Human Ecology: Method and Theory for a Contextual Approach.* Cambridge: Cambridge University Press.

Campbell, D. E.

1986   *Design and Operation of Irrigation Systems for Smallholder Agriculture in South Asia.* Rome: Food and Agriculture Organization of the United Nations, Investment Centre Technical Paper 3/1.

Carlstein, Tommy

1982   *Time, Resources, Society and Ecology: On the Capacity for Human Interaction in Space and Time,* vol. 1, *Preindustrial Societies.* London: George Allen and Unwin.

Carruthers, Ian, and Colin Clark

1981   *The Economics of Irrigation.* Liverpool: Liverpool University Press.

Caso, Alfonso

1937   *The Religion of the Aztecs.* Mexico City: Popular Library of Mexican Culture.

Chapultepec: Historia y presencia

1988   Mexico City: Smufit Cartón y Papel de México, SA de CV.

Chardon, Roland

1980   The Linear League in North America. *Annals of the Association of American Geographers* 70:129–153.

Charlton, Thomas H.

1977   Report on a Prehispanic Canal System, Otumba, Edo. de Mexico: Archaeological Investigations, August 10–19, 1977. Informe al Instituto Nacional de Antropología e Historia, Mexico City.

1978   Investigaciones arqueológicas en el municipio de Otumba, primera parte: Resultos preliminares de los trabajos de campo, 1978. Informe al Consejo de Arqueología, Instituto Nacional de Antropología e Historia, Mexico City.

1979a   Investigaciones arqueológicas en el municipio de Otumba, segunde parte: La Cerámica. Informe al Consejo de Arqueología, Instituto Nacional de Antropología e Historia, Mexico City.

1979b   Investigaciones arqueológicas en el municipio de Otumba, quinto parte, El riego y el intercambio: La expansión de Tula. Informe Final al Consejo de Arqueología, Instituto Nacional de Antropología e Historia, Mexico City.

Coe, Michael D.

1964   The Chinampas of Mexico. *Scientific American* 211:90–98.

1968   San Lorenzo and the Olmec Civilization. In *Dumbarton Oaks Conference on the Olmecs,* edited by Elizabeth P. Benson, pp. 41–77. Washington, D.C.: Dumbarton Oaks Research Library and Collection.

1981   San Lorenzo Tenochtitlan. In *Supplement to the Handbook of Middle American Indians*, vol. 1, *Archaeology*, edited by Victoria Reifler Bricker and Jeremy A. Sabloff, pp. 117–146. Austin: University of Texas Press.

Coe, Michael D., and Richard A. Diehl
1980   *In the Land of the Olmec*, vol. 1, *The Archaeology of San Lorenzo Tenochtitlán*. Austin: University of Texas Press.

Crespo Oviedo, Ana María
1976   Uso del suelo y patrón de poblamiento en el área de Tula, Hgo. In *Proyecto Tula, segunda parte*, coordinated by Eduardo Matos Moctezuma, pp. 35–48. Mexico City: Instituto Nacional de Antropología e Historia, Colección Científica 33.

Crossley, Mimi
1986   Mexico Stone Find Is North America's Oldest. *Austin American-Statesman*, 27 April.

Crosswhite, Frank S.
1981–1982   Corn (Zea mays) in Relation to Its Wild Relatives. *Desert Plants* 3:193–202.

Crumrine, N. Ross
1983   Mayo. In *Handbook of North American Indians*, vol. 10, *Southwest*, edited by Alfonso Ortiz, pp. 265–275. Washington, D.C.: Smithsonian Institution.

Darch, J. P., ed.
1983   *Drained Field Agriculture in Central and South America*. Oxford: B.A.R. International Series 189.

Daumas, Maurice, ed.
1962–1968   *Histoire générale des techniques.* 3 vols. Paris: Presses Universitaires de France.

Dávila Bonilla, Fray Agustín
1625   *Historia de la fundación y discurso de la provincia de Santiago de México, de la Orden de Preciadores.* Brussels: Ivan de Meerbeque.

Denevan, William M.
1983   Adaptation, Variation and Cultural Geography. *Professional Geographer* 35:399–407.

Díaz, Bernal
1963   *The Conquest of New Spain.* New York: Penguin Books.

Diehl, Richard A.
1974   Summary and Conclusions. In *Studies of Ancient Tollan: A Report of the University of Missouri Tula Archaeological Project*, edited by Richard A. Diehl, pp. 190–195. Columbia: University of Missouri Monographs in Anthropology 1.

1976   Pre-Hispanic Relationships between the Basin of Mexico and North and West Mexico. In *The Valley of Mexico: Studies in Pre-Hispanic Ecology and Society*, edited by Eric R. Wolf, pp. 249–286. Albuquerque: University of New Mexico Press for the School of American Research.

1981   Tula. In *Supplement to the Handbook of Middle American Indians,* edited by Victoria Reifler Bricker and Jeremy A. Sabloff, pp. 277–295. Austin: University of Texas Press.

1983   *Tula: The Toltec Capital of Ancient Mexico.* London: Thames and Hudson.

DiPeso, Charles C.

1966   Archaeology and Ethnohistory of the Northern Sierra. In *Handbook of Middle American Indians,* vol. 4, *Archaeological Frontiers and External Connections,* edited by Robert Wauchope, Gordon F. Ekholm, and Gordon R. Willey, pp. 3–25. Austin: University of Texas Press.

1974   *Casas Grandes: A Fallen Trading Center of the Gran Chichimeca,* vols. 1–3. Flagstaff: Northland Press for the Amerind Foundation Inc., Dragoon, Ariz.

1983   The Northern Sector of the Mesoamerican World System. In *Forgotten Places and Things: Archaeological Perspectives on American History,* edited by Albert E. Ward, pp. 11–22. Albuquerque: Center for Anthropological Studies.

1984   The Structure of the 11th Century Casas Grandes Agricultural System. In *Prehistoric Agricultural Strategies in the Southwest,* edited by Suzanne K. Fish and Paul R. Fish, pp. 261–269. Tempe: Arizona State University Anthropological Research Papers 33.

DiPeso, Charles C., John B. Renaldo, and Gloria J. Fenner

1974   *Casas Grandes: A Fallen Trading Center in the Gran Chichimeca,* vols. 4–8. Flagstaff: Northland Press for the Amerind Foundation Inc., Dragoon, Ariz.

Donkin, R. A.

1979   *Agricultural Terracing in the New World.* Viking Fund Publications in Anthropology 56. Tucson: University of Arizona Press for the Wenner-Gren Foundation for Anthropological Research, Inc.

Doolittle, William E.

1980   Aboriginal Agricultural Development in the Valley of Sonora, Mexico. *Geographical Review* 70: 328–342.

1984a   Agricultural Change as an Incremental Process. *Annals of the Association of American Geographers* 74: 124–138.

1984b   Cabeza de Vaca's Land of Maize: An Assessment of Its Agriculture. *Journal of Historical Geography* 10: 246–262.

1988   *Pre-Hispanic Occupance in the Valley of Sonora, Mexico: Archaeological Confirmation of Early Spanish Reports.* Anthropological Papers of the University of Arizona 48. Tucson: University of Arizona Press.

in press *Pocitos* and *Registros:* Comments on Water Control Features at Hierve el Agua, Oaxaca. *American Antiquity.*

Doyel, David E.
1979 The Prehistoric Hohokam of the Arizona Desert. *American Scientist* 67:544–554.

Drennan, Robert D., and Kent V. Flannery
1983 The Growth of Site Hierarchies in the Valley of Oaxaca: Part II. In *The Cloud People: Divergent Evolution of the Zapotec and Mixtec Civilizations*, edited by Kent V. Flannery and Joyce Marcus, pp. 65–70. New York: Academic Press.

Feldman, Lawrence H.
1974a Tollan in Hidalgo: Native Accounts of the Central Mexican Tolteca. In *Studies of Ancient Tollan: A Report of the University of Missouri Tula Archaeological Project*, edited by Richard A. Diehl, pp. 130–149. Columbia: University of Missouri Monographs in Anthropology 1.
1974b Tollan in Central Mexico: The Geography of Economic Specialization. In *Studies of Ancient Tollan: A Report of the University of Missouri Tula Archaeological Project*, edited by Richard A. Diehl, pp. 150–189, Columbia: University of Missouri Monographs in Anthropology 1.

Fernández Ordóñez, José A., Rosario Martínez Vázquez de Parga, Teresa Sánchez Lázaro, Luis de Carrera González, and Alejandro Carro Pérez
1986 *Catálogo de treinta canales españoles anteriores a 1900.* Madrid: Comisión de Estudios Históricos de Obras Públicas y Urbanismo.

Flannery, Kent V.
1970 Preliminary Archeological Investigations in the Valley of Oaxaca, Mexico, 1966 through 1969. Report to the National Science Foundation and the Instituto Nacional de Antropología e Historia.
1983 Precolumbian Farming in the Valleys of Oaxaca, Nochixtlan, Tehuacán, and Cuicatlán: A Comparative Study. In *The Cloud People: Divergent Evolution of the Zapotec and Mixtec Civilizations*, edited by Kent V. Flannery and Joyce Marcus, pp. 323–339. New York: Academic Press.

Flannery, Kent V., Anne V. T. Kirkby, Michael J. Kirkby, and Aubrey W. Williams, Jr.
1967 Farming Systems and Political Growth in Ancient Oaxaca. *Science* 158:445–454.

Flannery, Kent V., and Joyce Marcus
1976 Formative Oaxaca and the Zapotec Cosmos. *American Scientist* 64:374–383.

Ford, Richard I.
1977 The Technology of Irrigation in a New Mexico Pueblo. In *Material Culture: Styles, Organization, and Dynamics of Technology*, edited by Heather Lechtman and Robert S. Merrill, pp. 134–154. St. Paul: West Publishing Co.

Fowler, Melvin L.
   1969   A Preclassic Water Distribution System in Amalucan, Mexico.
            Archaeology 22:208–215.
   1987   Early Water Management at Amalucan, State of Puebla, Mexico.
            National Geographic Research 3:52–68.
French, N., and I. Hussain
   1964   Water Spreading Manual. Lahore, Pakistan: West Pakistan Range
            Improvement Scheme, Range Management Record 1.
Fuentes Aguilar, Luis
   1988   Irrigación y urbanismo en la Cuenca Mesoamericana de México.
            Revista Geográfica 107:5–28.
García Cook, Angel
   1981   The Historical Importance of Tlaxcala in the Cultural Develop-
            ment of the Central Highlands. In Supplement to the Hand-
            book of Middle American Indians, vol. 1, Archaeology, edited
            by Victoria Reifler Bricker and Jeremy A. Sabloff, pp. 244–295.
            Austin: University of Texas Press.
   1985   Historia de la tecnología agrícola en el Altiplano Central desde el
            principio de la agricultura hasta el siglo XIII. In Historia de la
            Agricultura Epoca Prehispánica Siglo XVI, edited by Teresa
            Rojas Rabiela and William T. Sanders, vol. 2, pp. 7–77. Mexico
            City: Instituto Nacional de Antropología e Historia.
   1986   El control de la erosión en Tlaxcala: Un problema secular. Er-
            kunde 40:251–262.
García-Diego, José A.
   1977   Old Dams in Extremadura. In History of Technology, edited by
            A. Rupert Hall and Norman Smith, vol. 2, pp. 95–124. London:
            Mansell.
García Payón, José
   1971   Archaeology of Central Veracruz. In Handbook of Middle Ameri-
            can Indians, vol. 11, Archaeology of Northern Mesoamerica,
            Part 2, edited by Robert Wauchope, Gordon F. Ekholm, and
            Ignacio Bernal, pp. 505–542. Austin: University of Texas Press.
Gerhard, Peter
   1982   The Northern Frontier of New Spain. Princeton: Princeton Uni-
            versity Press.
Gil, Gorgonio, and James A. Neely
   1967   Historia de la fundación del pueblo de San Gabriel Chilacatla.
            Tlalocan 5:198–219.
Golomb, Berl, and Herbert M. Eder
   1964   Landforms Made by Man. Landscape 1:4–7.
Grove, David C.
   1984   Chalcatzingo: Excavations on the Olmec Frontier. London:
            Thames and Hudson.
Grove, David C., and Ann Cyphers Guillén
   1987   The Excavations. In Ancient Chalcatzingo, edited by David C.
            Grove, pp. 21–55. Austin: University of Texas Press.

Grove, David C., Kenneth G. Hirth, David E. Bugé, and Ann M. Cyphers
1976   Settlement and Cultural Development at Chalcatzingo. *Science*
192:1203–1210.
Hammond, George P., and Agapito Rey, trans., eds., and annots.
1928   *Obregon's History of 16th Century Explorations in Western
America Entitled: Chronicle, Commentary, or Relation of the
Ancient and Modern Discoveries in New Spain and New Mexico, Mexico, 1584.* Los Angeles: Wetzel.
1929   *Expedition into New Mexico Made by Antonio de Espejo, 1582–
1583: As Revealed in the Journal of Diego Pérez de Luxán, a
Member of the Party.* Los Angeles: Quivira Society.
Haury, Emil W.
1976   *The Hohokam Desert Farmers and Craftsmen: Excavations at
Snaketown, 1964–1965.* Tucson: University of Arizona Press.
Heizer, Robert F., and James A. Bennyhoff
1958   Archaeological Investigation of Cuicuilco, Valley of Mexico, 1957.
Science 127:232–233.
Herold, Laurance C.
1965   *Trincheras and Physical Environment along the Rio Gavilan,
Chihuahua, Mexico.* Denver: University of Denver, Department of Geography, Publications in Geography, Technical
Paper 65-1.
Hewitt, William P., Marcus C. Winter, and David A. Peterson
1987   Salt Production at Hierve el Agua, Oaxaca. *American Antiquity*
52:799–814.
Hillel, Daniel
1987   *The Efficient Use of Water in Irrigation: Principles and Practices
for Improving Irrigation in Arid and Semiarid Regions.* Washington, D.C.: World Bank, Technical Paper No. 64.
Hinton, Thomas
1983   Southern Periphery: West. In *Handbook of North American Indians,* vol. 10, *Southwest,* edited by Alfonso Ortiz, pp. 315–328.
Washington, D.C.: Smithsonian Institution.
Hopkins, Joseph W., III
1968   Prehispanic Agricultural Terraces in Mexico. M.A. thesis, Department of Anthropology, University of Chicago.
1983   The Tomellín Cañada and the Postclassic Cuicatec. In *The Cloud
People: Divergent Evolution of the Zapotec and Mixtec Civilizations,* edited by Kent V. Flannery and Joyce Marcus, pp. 266–
270. New York: Academic Press.
1984   *Irrigation and the Cuicatec Ecosystem: A Study of Agriculture
and Civilization in North Central Oaxaca.* Ann Arbor: Memoirs of the Museum of Anthropology, University of Michigan, 17.
Howard, William A., and Thomas M. Griffiths
1966   *Trinchera Distribution in the Sierra Madre Occidental, Mexico.*

Denver: University of Denver, Department of Geography, Publications in Geography, Technical Paper 66-1.

Hunt, Eva

1972   Irrigation and the Socio-Political Organization of the Cuicatec Cacicazgos. In *The Prehistory of the Tehuacan Valley*, vol. 4, *Chronology and Irrigation*, edited by Richard S. MacNeish and Frederick Johnson, pp. 162–259. Austin and London: University of Texas Press for the Robert S. Peabody Foundation.

Hunt, Eva, and Robert C. Hunt

1974   Irrigation, Conflict, and Politics: A Mexican Case. In *Irrigation's Impact on Society*, edited by Theodore E. Downing and McGuire Gibson, pp. 129–157. Anthropological Papers of the University of Arizona 25. Tucson: University of Arizona Press.

Hunt, Robert C.

1988   Size and the Structure of Authority in Canal Irrigation Systems. *Journal of Anthropological Research* 44:335–355.

International Land Development Consultants

1981   *Agricultural Compendium for Rural Development in the Tropics and Subtropics.* Amsterdam: Elsevier Scientific Publishing Co.

Israelson, Orson W., and Vaughn E. Hansen

1962   *Irrigation Principles and Practices.* New York: John Wiley and Sons.

Ixtlilxóchitl, Fernando de Alva

1891   *Obras históricas.* 2 vols. Mexico City: Oficina de la Secretaría de Fomento.

Jansen, P. Ph., K. van Bendegom, J. van den Berg, M. de Vries, and A. Zanen, eds.

1979   *Principles of River Engineering: The Non-Tidal Alluvial River.* London: Pitman Publishers.

Kappel, Wayne

1974   Irrigation Development and Population Pressure. In *Irrigation's Impact on Society*, edited by Theodore E. Downing and McGuire Gibson, pp. 159–167. Anthropological Papers of the University of Arizona 25. Tucson: University of Arizona Press.

Kelley, J. Charles

1949a   Archaeological Notes on Two Excavated House Structures in Western Texas. *Texas Archaeological and Paleontological Society, Bulletin* 20:89–114.

1949b   Notes on Julimes, Chihuahua. *El Palacio* 56:358–361.

1952a   Factors Involved in the Abandonment of Certain Peripheral Southwestern Settlements. *American Anthropologist* 54:356–387.

1952b   The Historic Indian Pueblos of La Junta de los Rios. *New Mexico Historical Review* 27:257–295.

1953   The Historic Indian Pueblos of La Junta de los Rios. *New Mexico Historical Review* 28:21–51.

1956   Settlement Patterns in North-central Mexico. In *Prehistoric Settlement Patterns in the New World*, edited by Gordon R.

Willey, pp. 128–139. New York: Viking Fund Publications in Anthropology 23.

1966 Mesoamerica and the Southwestern United States. In *Handbook of Middle American Indians*, vol. 4, *Archaeological Frontiers and External Connections*, edited by Robert Wauchope, Gordon F. Ekholm, and Gordon R. Willey, pp. 95–110. Austin: University of Texas Press.

1971 Archaeology of the Northern Frontier: Zacatecas and Durango. In *Handbook of Middle American Indians*, vol. 11, *Archaeology of Northern Mesoamerica, Part 2*, edited by Robert Wauchope, Gordon F. Ekholm, and Ignacio Bernal, pp. 768–801. Austin: University of Texas Press.

1986 *Jumano and Patarabueye: Relations at La Junta de los Rios*. Ann Arbor: Anthropological Papers, Museum of Anthropology, University of Michigan, 77.

Kelley, J. Charles, and William J. Shackelford

1954 Preliminary Notes on the Weicker Site, Durango, Mexico. *El Palacio* May: 145–150.

Kelly, Isabel, and Angel Palerm

1952 *The Tajín Totonac, Part 1: History, Subsistence, Shelter and Technology*. Washington, D.C.: Smithsonian Institution, Institute of Social Anthropology, Publication 13.

Kern, Horst

1973 Estudios geográficos sobre residuos de poblados y campos en el Valle de Puebla-Tlaxcala. *Comunicaciones Proyecto Puebla-Tlaxcala* 7: 73–75.

Kinney, Clesson S.

1918 History of Ancient Irrigation in Various Countries. *Irrigation Age* 33: 86–89.

Kirch, Patrick V.

1980 The Archaeological Study of Adaptation: Theoretical and Methodological Issues. In *Advances in Archaeological Method and Theory*, vol. 3, edited by Michael J. Schiffer, pp. 101–156. New York: Academic Press.

Kirkby, Anne V. T.

1973 *The Use of Land and Water Resources in the Past and Present Valley of Oaxaca, Mexico*. Ann Arbor: Memoirs of the Museum of Anthropology, University of Michigan, 5.

Krantz, B. A., and J. Kampen

1979 Crop Production Systems in Semi-Arid Tropical Zones. In *Soil, Water and Crop Production*, edited by D. Wynne Thorne and Marlowe D. Thorne, pp. 278–298. Westport, Conn.: AVI Publishing Co.

Kranzberg, Melvin, and Carroll W. Pursell, Jr.

1967 *Technology in Western Civilization*. 2 vols. New York: Oxford University Press.

Kroeber, A. L.
  1940   Stimulus Diffusion. *American Anthropologist* 42:1–20.
  1953   *Cultural and Natural Areas of North America.* Berkeley and Los Angeles: University of California Press.
Kubler, George
  1944   Architects and Builders in Mexico, 1521–1550. *Journal of the Warburg and Courtland Institutes* 7:7–19.
Lameiras, José
  1974   Relación en torno a la posesión de tierras y aguas: Un pleito entre indios principales de Teotihuacan y Acolman en el siglo XVI. In *Nuevas noticias sobre las obras hidráulicas: Prehispánicas y coloniales en el Valle de México,* edited by Teresa Rojas R., Rafael Strauss K., and José Lameiras, pp. 177–199. Mexico City: Instituto Nacional de Antropología e Historia.
Lauer, Wilhelm
  1973   Zusammenhänge zwischen Klima und Vegetation am Ostabfall der Mexikanischen Meseta. *Erkunde* 27:192–213
Lawton, H. W., and P. J. Wilke
  1979   Ancient Agricultural Systems in Dry Regions. In *Agriculture in Semi-Arid Environments,* edited by A. E. Hall, G. H. Cannell, and H. W. Lawton, pp. 1–43. Berlin and New York: Springer-Verlag.
Layton, Edward T., Jr.
  1974   Technology as Knowledge. *Technology and Culture* 15:31–41.
Lees, Susan H.
  1970   Test Excavations at Santo Domingo Tomaltepec (B-176). In Preliminary Archaeological Investigations in the Valley of Oaxaca, Mexico, 1966 through 1969, by Kent V. Flannery, pp. 80–82. Report to the National Science Foundation and the Instituto Nacional de Antropología e Historia.
  1973   *Sociopolitical Aspects of Canal Irrigation in the Valley of Oaxaca.* Ann Arbor: Memoirs of the Museum of Anthropology, University of Michigan, 6.
Lekson, Stephen H.
  1984   Dating Casas Grandes. *Kiva* 50:55–60.
Lister, Robert H.
  1978   Mesoamerican Influence at Chaco Canyon, New Mexico. In *Across the Chichimec Sea: Papers in Honor of J. Charles Kelley,* edited by Carroll L. Riley and Basil C. Hedrick, pp. 233–241. Carbondale and Edwardsville: Southern Illinois University Press.
López de Gómara, Francisco
  1943   Historia de la conquista de México. Mexico City: Editorial Pedro Robredo.
Lorenzo, José Luis
  1968   Clima y agricultura en Teotihuacán. In *Materiales para la arqueología de Teotihuacán,* edited by José Luis Lorenzo, pp. 51–

72. Mexico City: Instituto Nacional de Antropología e Historia, Serie Investigaciones 17.

Los Anales de Cuauhtitlán

1945　*Códice Chimalpopoca*. Mexico City: Universidad Nacional Autónoma de México.

McAfee, Byron, and R. H. Barlow, trans. and annot.

1946　The Titles of Tetzcotzinco (Santa María Nativitas). *Tlalocan* 2:110–127.

McGuire, Randall H.

1980　The Mesoamerican Connection in the Southwest. *Kiva* 46:3–38.

MacNeish, Richard S.

1962　*Second Annual Report of the Tehuacán Archaeological-Botanical Project.* Andover, Mass.: Robert S. Peabody Foundation.

1964a　Ancient Mesoamerican Civilization. *Science* 143:531–537.

1964b　The Origins of New World Civilization. *Scientific American* 211:3–11.

1971　Speculations about How and Why Food Production and Village Life Developed in the Tehuacan Valley, Mexico. *Archaeology* 24:307–315.

1972　The Evolution of Community Patterns in the Tehuacán Valley of Mexico and Speculations about the Cultural Processes. In *Man, Settlement and Urbanism*, ed. Peter J. Ucko, Ruth Tringham, and G. W. Dimbleby, pp. 67–93. Hertfordshire: Duckworth.

Mangelsdorf, Paul C.

1974　*Corn: Its Origin, Evolution, and Improvement.* Cambridge: Belknap Press of Harvard University Press.

Manrique C., Leonardo

1969　The Otomi. In *Handbook of Middle American Indians*, vol. 8, *Ethnology, Part 2*, edited by Robert Wauchope and Evon Z. Vogt, pp. 682–722. Austin: University of Texas Press.

Martínez Donjuán, Guadalupe

1986　Teopantecuanitlán. In *Arqueología y etnohistoria del estado de Guerrero*, pp. 55–80. Mexico City: Instituto Nacional de Antropología e Historia; Gobierno del Estado de Guerrero.

Mason, J. Alden

1937　Late Archaeological Sites in Durango, Mexico: From Chalchihuites to Zape. In *Publications of the Philadelphia Anthropological Society*, vol. 1, *Twenty-fifth Anniversary Studies*, edited by D. S. Davidson, pp. 127–146. Philadelphia: University of Pennsylvania Press.

Mason, Roger D., Dennis E. Lewarch, Michael J. O'Brien, and James A. Neely

1977　An Archaeological Survey of the Xoxocotlan Piedmont, Oaxaca, Mexico. *American Antiquity* 42:567–575.

Masse, W. Bruce

1987a　Physical and Cultural Setting. In *Archaeological Investigations of Portions of the Las Acequias–Los Muertos Irrigation Sys-*

tem: Testing and Partial Data Recovery within the Tempe Section of the Outer Loop Freeway System, Maricopa County, Arizona, edited by W. Bruce Masse, pp. 11–16. Tucson: Arizona State Museum, Archaeological Series 176.

1987b   Syntheses and Discussion. In *Archaeological Investigations of Portions of the Las Acequias–Los Muertos Irrigation System: Testing and Partial Data Recovery within the Tempe Section of the Outer Loop Freeway System, Maricopa County, Arizona,* edited by W. Bruce Masse, pp. 189–196. Tucson: Arizona State Museum, Archaeological Series 176.

Masse, W. Bruce, and Robert W. Layhe

1987   Testing and Data Recovery at La Cuenca del Sedimento, AZ U:9:68 (ASM). In *Archaeological Investigations of Portions of the Las Acequias–Los Muertos Irrigation System: Testing and Partial Data Recovery within the Tempe Section of the Outer Loop Freeway System, Maricopa County, Arizona,* edited by W. Bruce Masse, pp. 83–122. Tucson: Arizona State Museum, Archaeological Series 176.

Mastache de Escobar, Alba Guadalupe

1976   Sistemas de riego en el área de Tula, Hgo., In *Proyecto Tula, segunda parte,* coordinated by Eduardo Matos Moctezuma, pp. 49–70. Mexico City: Instituto Nacional de Antropología e Historia, Colección Científica 33.

Matheny, Raymond T.

1982   Ancient Lowland and Highland Maya Water and Soil Conservation Strategies. In *Maya Subsistence: Studies in Memory of Dennis E. Puleston,* edited by Kent V. Flannery, pp. 157–178. New York: Academic Press.

Matheny, Ray T., and Deanne L. Gurr

1979   Ancient Hydraulic Techniques in the Chiapas Highlands. *American Scientist* 67:441–449.

1983   Variations in Prehistoric Agricultural Systems of the New World. *Annual Review of Anthropology* 12:79–103.

Mendizábal, Miguel Othón de

1946   El jardin de Netzahualcoyotl en el Cerro de Tetzcotzinco. In *Obras completas,* vol. 2, pp. 443–451. Mexico City: Talleres Gráficos de la Nación.

Meyer, Michael C.

1984   *Water in the Hispanic Southwest: A Social and Legal History, 1550–1850.* Tucson: University of Arizona Press.

Millon, René F.

1954   Irrigation at Teotihuacan. *American Antiquity* 20:177–180.

1957   Irrigation Systems in the Valley of Teotihuacan. *American Antiquity* 23:160–166.

Monkhouse, F. J.

1965   *A Dictionary of Geography.* Chicago: Aldine Publishing Co.

Moore, W. G.
1974   *A Dictionary of Geography: Definitions and Explanations of Terms Used in Physical Geography,* 5th ed. Middlesex: Penguin Books.

Morisawa, Marie
1968   *Streams: Their Dynamics and Morphology.* New York: McGraw-Hill.

Morris, F. Bayard, trans.
1928   *Hernando Cortés: Five Letters, 1519–1526.* London: George Routledge and Sons.

Multhauf, Robert P.
1959   The Scientist and the 'Improver' of Technology. *Technology and Culture* 1:38–47.

Murphy, Michael E.
1986   *Irrigation in the Bajío Region of Colonial Mexico.* Boulder and London: Westview Press, Dellplain Latin American Studies 19.

Nabhan, Gary Paul
1979   The Ecology of Floodwater Farming in Arid Southwestern North America. *Agro-Ecosystems* 5:245–255.
1986a   Papago Indian Desert Agriculture and Water Control in the Sonoran Desert, 1697–1934. *Applied Geography* 6:43–59.
1986b   '*AK-ciñ* "Arroyo Mouth" and the Environmental Setting of the Papago Indian Fields in the Sonoran Desert. *Applied Geography* 6:61–75.

Naylor, Thomas H., and Charles W. Polzer, comps. and eds.
1986   *The Presidio and Militia on the Northern Frontier of New Spain— A Documentary History, Volume One: 1570–1700.* Tucson: University of Arizona Press.

Neely, James A.
1967   Organización hidráulica y sistemas de irrigación prehistóricos en el Valle de Oaxaca. *Boletín del INAH* 27:15–17.
1970   Terrace and Water Control Systems in the Valley of Oaxaca Region: A Preliminary Report. In Preliminary Archaeological Investigations in the Valley of Oaxaca, Mexico, 1966 through 1969, by Kent V. Flannery, pp. 83–87. Report to the National Science Foundation and the Instituto Nacional de Antropología e Historia.

Nicholas, Linda, and Jill Neitzel
1984   Canal Irrigation and Sociopolitical Organization in the Lower Salt River Valley: A Diachronic Analysis. In *Prehistoric Agricultural Strategies in the Southwest,* edited by Suzanne K. Fish and Paul R. Fish, pp. 161–178. Tempe: Arizona State University Anthropological Research Papers 33.

Nichols, Deborah L.
1982a   A Middle Formative Irrigation System near Santa Clara Coatitlan in the Basin of Mexico. *American Antiquity* 47:133–143.
1982b   Results of the Excavations of the Tlajinga Canals. In *A Recon-*

struction of a Classic Period Landscape in the Teotihuacan
Valley: Final Report to the National Science Foundation, edited
by William T. Sanders, Deborah Nichols, Rebecca Storey, and
Randolph Widner, pp. 99–127. University Park: Department of
Anthropology, Pennsylvania State University.

1988   Infrared Aerial Photography and Prehispanic Irrigation at Teotihuacán: The Tlajinga Canals. *Journal of Field Archaeology*
15:17–27.

Niekler, Otto

1919   Texcotzinco. *México Antiguo* 1:110–112.

Nir, Dov

1983   *Man: A Geomorphic Agent.* Jerusalem: Keter Publishing House.

O'Brien, Michael J., Dennis E. Lewarch, Roger D. Mason, and James A.
Neely

1980   Functional Analysis of Water Control Features at Monte Alban,
Oaxaca, Mexico. *World Archaeology* 11:342–355.

O'Brien, Michael J., Roger D. Mason, Dennis E. Lewarch, and James A.
Neely

1982   *A Late Formative Irrigation Settlement below Monte Albán: Survey and Excavation on the Xoxocotlán Piedmont, Oaxaca,
Mexico.* Austin: University of Texas Press.

Olin, W. H.

1913   *American Irrigation Farming.* Chicago: A. C. McClurg and Co.

Ortloff, Charles R., Michael E. Moseley, and Robert A. Feldman

1982   Hydraulic Engineering Aspects of the Chimu Chicama-Moche
Intervalley Canal. *American Antiquity* 47:572–595.

Palerm, Angel

1954   La distribución de regadío en al área central de Mesoamérica.
Ciencias Sociales 5:2–15, 64–74.

1955   The Agricultural Bases of Urban Civilization in Mesoamerica. In
Irrigation Civilizations: A Comparative Study, by Julian H.
Steward et al., pp. 28–42. Washington, D.C.: Pan American
Union, Social Science Monographs 1.

1961a   Distribución del regadío prehispánico en el área central de
Mesoamérica. *Revista Interamericana de Ciencias Sociales*
1:242–267.

1961b   Sistemas de regadío en Teotihuacán y en el Pedregal. *Revista
Interamericana de Ciencias Sociales* 1:297–302.

1967   Agricultural Systems and Food Patterns. In *Handbook of Middle
American Indians,* vol. 6, *Social Anthropology,* edited by Robert
Wauchope and Manning Nash, pp. 26–52. Austin: University
of Texas Press.

1973   *Obras hidráulicas prehispánicas en el sistema lacustre del Valle
de México.* Mexico City: Instituto Nacional de Antropología e
Historia.

Palerm, Angel, and Eric Wolf
1961   Sistemas agrícolas y desarrollo del área clave del imperio Texcoca-
        no. *Revista Interamericana de Ciencias Sociales* 1:281–287.
Parsons, Jeffrey R.
1971   *Prehistoric Settlement Patterns in the Texcoco Region, Mexico.*
        Ann Arbor: Memoirs of the Museum of Anthropology, Univer-
        sity of Michigan, 3.
Peña C., Agustín, and Carmen Rodríguez
1976   Excavaciones en Daini, Tula, Hgo. In *Proyecto Tula, segunda
        parte,* coordinated by Eduardo Matos Moctezuma, pp. 85–90.
        Mexico City: Instituto Nacional de Antropología e Historia,
        Colección Científica 33.
Pennington, Campbell W.
1963   *The Tarahumar of Mexico: Their Environment and Material Cul-
        ture.* Salt Lake City: University of Utah Press.
1969   *The Tepehuan of Chihuahua: Their Material Culture.* Salt Lake
        City: University of Utah Press.
1980   *The Pima Bajo of Central Sonora, Mexico,* vol. 1, *The Material
        Culture.* Salt Lake City: University of Utah Press.
Pomar, Juan Bautista
1941   *Relaciones de Texcoco y de la Nueva España.* Mexico City:
        S. Chávez Hayhoe.
Porter Weaver, Muriel
1972   *The Aztecs, Maya, and Their Predecessors: Archaeology of Meso-
        america.* New York: Seminar Press.
Prager, Frank D.
1978   Vitruvius and the Elevated Aqueducts. In *History of Technol-
        ogy,* vol. 3, edited by A. Rupert Hall and Norman Smith,
        pp. 105–121. London: Mansell.
Precourt, Prudence S.
1983   Settlements, Systems, and Patterns: An Ecological Systems
        Analysis of Settlement Systems near Amozoc de Mota, Puebla,
        Mexico. Ph.D. Dissertation, Department of Anthropology, Uni-
        versity of Wisconsin–Milwaukee.
Prem, Hanns J.
1975   Los afluentes del Río Xopanac: Estudio histórico de un sistema de
        riego. *Comunicaciones Proyecto Puebla-Tlaxcala* 12:27–40.
Price, Barbara J.
1971   Prehispanic Irrigation Agriculture in Nuclear America. *Latin
        American Research Review* 6.3:3–60.
Prindiville, Mary, and David C. Grove
1987   The Settlement and Its Architecture. In *Ancient Chalcatzingo,*
        edited by David C. Grove, pp. 63–81. Austin: University of
        Texas Press.
Rawitz, E.
1973   Gravity Irrigation. In *Arid Zone Irrigation,* edited by B. Yaron,

E. Danfors, and Y. Vaadia, pp. 323–337. Heidelberg and New York: Springer-Verlag.

Redmond, Elsa M.
1983   A fuego y sangre: Early Zapotec Imperialism in the Cuicatlán Cañada, Oaxaca. Ann Arbor: Memoirs of the Museum of Anthropology, University of Michigan, 16.

Redmond, Elsa M., and Charles S. Spencer
1983   The Cuicatlán Cañada and the Period II Frontier Zapotec State. In The Cloud People: Divergent Evolution of the Zapotec and Mixtec Civilizations, edited by Kent V. Flannery and Joyce C. Marcus, pp. 117–120. New York: Academic Press.

Richards, Keith
1982   Rivers: Form and Process in Alluvial Channels. London: Methuen.

Rindos, David
1984   The Origins of Agriculture: An Evolutionary Perspective. Orlando: Academic Press.

Rojas Rabiela, Teresa
1974   Aspectos tecnológicos de las obras hidráulicas coloniales. In Nuevas noticias sobre las obras hidráulicas prehispánicas y coloniales en el Valle de México, edited by Teresa Rojas Rabiela, Rafael A. Strauss K., and José Lameiras, pp. 19–133. Mexico City: Instituto Nacional de Antropología e Historia.
1985   La tecnología agrícola mesoamericana en el siglo XVI. In Historia de la agricultura, época prehispánica, siglo XVI, vol. 1, edited by Teresa Rojas Rabiela and William T. Sanders, pp. 129–214. Mexico City: Instituto Nacional de Antropología e Historia.
1988   Las siembras de ayer: La agricultura indígena del siglo XVI. Mexico City: SEP.

Rojas Rabiela, Teresa, Rafael A. Strauss K., and José Lameiras, eds.
1974   Nuevas noticias sobre las obras hidráulicas prehispánicas y coloniales en el Valle de México. Mexico City: Instituto Nacional de Antropología e Historia.

Romero de Terreros, Manuel
1949   Los acueductos de México en la historia y en el arte. Mexico City: Universidad Nacional Autónoma de México, Instituto de Investigaciones Estéticas.

Rydzewski, J. R., ed.
1987   Irrigation Development Planning: An Introduction for Engineers. Chichester and New York: John Wiley and Sons.

Sanders, William T.
1956   The Central American Symbiotic Region: A Study in Prehistoric Settlement Patterns. In Prehistoric Settlement Patterns in the New World, edited by Gordon R. Willey, pp. 115–127. New York: Viking Fund Publications in Anthropology 23.
1965   The Cultural Ecology of the Teotihuacan Valley: A Preliminary Report of the Results of the Teotihuacan Valley Project. Uni-

versity Park: Department of Anthropology, Pennsylvania State University.

1971   Cultural Ecology and Settlement Patterns of the Gulf Coast. In *Handbook of Middle American Indians*, vol. 11, *Archaeology of Northern Mesoamerica, Part 2*, edited by Robert Wauchope, Gordon F. Ekholm, and Ignacio Bernal, pp. 543–557. Austin: University of Texas Press.

1976   The Agricultural History of the Basin of Mexico. In *The Valley of Mexico: Studies in Pre-Hispanic Ecology and Society*, edited by Eric R. Wolf, pp. 101–159. Albuquerque: University of New Mexico Press.

1981   Ecological Adaptation in the Basin of Mexico: 23,000 B.C. to the Present. In *Supplement to the Handbook of Middle American Indians*, vol. 1, *Archaeology*, edited by Victoria Reifler Bricker and Jeremy A. Sabloff, pp. 147–197. Austin: University of Texas Press.

1982   Introduction. In *A Reconstruction of a Classic Period Landscape in the Teotihuacan Valley: Final Report to the National Science Foundation*, edited by William T. Sanders, Deborah Nichols, Rebecca Storey, and Randolph Widmer, pp. 1–20. University Park: Department of Anthropology, Pennsylvania State University.

Sanders, William T., Deborah Nichols, Rebecca Storey, and Randolph Widmer, eds.

1982   *A Reconstruction of a Classic Period Landscape in the Teotihuacan Valley: Final Report to the National Science Foundation*. University Park: Department of Anthropology, Pennsylvania State University.

Sanders, William T., Jeffrey R. Parsons, and Robert S. Santley

1979   *The Basin of Mexico: Ecological Processes in the Evolution of a Civilization*. New York: Academic Press.

Sanders, William T., and Barbara J. Price

1968   *Mesoamerica: The Evolution of a Civilization*. New York: Random House.

Sanders, William T., and Robert S. Santley

1977   A Prehispanic Irrigation System near Santa Clara Xalostoc in the Basin of Mexico. *American Antiquity* 42:582–588.

SARH

1979   *La ingeniería civil en el desarrollo agropecuario de México*. Mexico City: Secretaría de Agricultura y Recursos Hidráulicos.

Sauer, Carl O.

1952   *Agricultural Origins and Dispersals: The Domestication of Animals and Foodstuffs*. New York: American Geographical Society.

1954   Comment on Gatherers and Farmers in the Greater Southwest: A Problem in Classification. *American Anthropologist* 56:553–556.

Sauer, Carol O., and Donald D. Brand
    1932   Aztatlán: Prehistoric Mexican Frontier on the Pacific Coast.
            Ibero-Americana 1.
Sayles, E. B.
    1936   An Archaeological Survey of Chihuahua, Mexico. Medallion
            Papers 22. Gila Pueblo.
Schiffer, Michael B.
    1986   Radiocarbon Dating and the "Old Wood" Problem: The Case of
            the Hohokam Chronology. Journal of Archaeological Science
            13:13–30.
Schmidt, Robert H., Jr., and Rex E. Gerald
    1988   The Distribution of Conservation-Type Water-Control Systems
            in the Northern Sierra Madre Occidental. Kiva 53:165–179.
Schroeder, Albert H.
    1965   Unregulated Diffusion from Mexico into the Southwest prior to
            A.D. 700. American Antiquity 30:297–309.
    1966   Pattern Diffusion from Mexico into the Southwest after A.D. 600.
            American Antiquity 31:683–704.
Seele, Enno
    1973   Restos de milpas y poblaciones prehispánicas cerca de San Buena-
            ventura Nealtican, Pue. Comunicaciones Proyecto Puebla-
            Tlaxcala 7:77–86.
Septien y Septien, Manuel
    1988   Acueducto y fuentes de Queretaro. Queretaro: Dirección de Pa-
            trimonio Cultural, Secretaría de Cultura y Bienestar Social,
            Gobierno del Estado de Queretaro, Colección Documentos 10.
Siemens, Alfred H.
    1983   Wetland Agriculture in Pre-Hispanic Mesoamerica. Geographical
            Review 73:166–181.
    1987   Review of Prehistoric Lowland Maya Environment and Subsis-
            tence Economy (Mary Pohl, ed.). Annals of the Association of
            American Geographers 77:666–668.
    n.d.   Modelling Pre-Hispanic Hydroagriculture on Levee Backslopes in
            Northern Veracruz. Manuscript.
Singer, Charles, E. J. Holmyard, and A. R. Hall
    1954–1958   A History of Technology. 5 vols. Oxford: Clarendon Press.
Smith, C. Earle, Jr.
    1965   Agriculture, Tehuacan Valley. Fieldiana: Botany 31:53–100.
Smith, Norman
    1971   A History of Dams. London: Peter Davies.
    1975   Man and Water: A History of Hydro-Technology. New York:
            Charles Scribner's Sons.
Smith, Norman A. F.
    1976   Attitudes to Roman Engineering and the Question of the Inverted
            Siphon. In History of Technology, vol. 1, edited by A. Rupert
            Hall and Norman Smith, pp. 45–71. London: Mansell.

Spencer, Charles S.
1982   *The Cuicatlán Cañada and Monte Albán: A Study of Primary State Formation.* New York: Academic Press.
Spencer, J. E., and G. A. Hale
1961   The Origin, Nature, and Distribution of Agricultural Terracing. *Pacific Viewpoint* 2 : 1–40.
Spicer, Edward H.
1980   *The Yaquis: A Cultural History.* Tucson: University of Arizona Press.
Spinden, Herbert J.
1915   The Origin and Distribution of Agriculture in America. *Proceedings of the Nineteenth International Congress of Americanists,* edited by F. W. Hodge, pp. 269–277. Washington, D.C.
1917   The Invention and Spread of Agriculture in America. *American Museum Journal* 17 : 181–188.
1928   *Ancient Civilizations of Mexico and Central America.* New York: American Museum of Natural History.
Steward, Julian H.
1929   Irrigation without Agriculture. *Papers of the Michigan Academy of Science, Arts and Letters* 12 : 149–156.
1977   Wittfogel's Irrigation Hypothesis. In *Evolution and Ecology: Essays on Social Transformation by Julian H. Steward,* edited by Jane C. Steward and Robert F. Murphy, 87–100. Urbana, Chicago, and London: University of Illinois Press.
Steward, Julian H., Robert M. Adams, Donald Collier, Angel Palerm, Karl A. Wittfogel, and Ralph L. Beals
1955   *Irrigation Civilizations: A Comparative Study; A Symposium on Method and Result in Cross-Cultural Regularities.* Washington, D.C.: Pan American Union, Social Science Monographs 1.
Strauss K., Rafael A.
1974   El área septentrional del Valle de México: Problemas agro hidráulicas, prehispánicos y coloniales. In *Nuevas noticias sobre las obras hidráulicas prehispánicas y coloniales en el Valle de México,* edited by Teresa Rojas Rabiela, Rafael A. Strauss K., and José Lameiras, pp. 137–174. Mexico City: Instituto Nacional de Antropología e Historia.
Tadmor, N. H., L. Shanan, and M. Evenari
1960   The Ancient Desert Agriculture of the Negev, VI: The Ratio of Catchment to Cultivated Area. *Ktavim* 10 : 193–221.
Taylor, William B.
1972   *Landlord and Peasant in Colonial Oaxaca.* Stanford: Stanford University Press.
Townsend, Richard Fraser
1982   Pyramid and Sacred Mountain. *Annals of the New York Academy of Sciences* 385 : 37–62.

Trautmann, Wolfgang
  1986   Geographical Aspects of Hispanic Colonization on the Northern Frontier of New Spain. *Erkurde* 40:241–250.
Trombold, Charles D.
  1976   Spatial Distribution, Functional Hierarchies, and Patterns of Interaction in Prehistoric Communities around La Quemada, Zacatecas, Mexico. In *Archaeological Frontiers: Papers on New World High Cultures in Honor of J. Charles Kelley,* edited by Robert B. Pickering, pp. 149–180. Carbondale: Southern Illinois University Museum Studies 4.
  1977   *The Role of Locational Analysis in the Development of Archaeological Research Strategy.* Ph.D. dissertation, Department of Anthropology, Southern Illinois University.
  1985   A Summary of the Archaeology in the La Quemada Region. In *The Archaeology of West and Northwest Mesoamerica,* edited by Michael S. Foster and Phil C. Wiegand, pp. 237–267. Boulder and London: Westview Press.
Turner, B. L., II
  1980   La agricultura intensiva de trabajo en las tierras mayas. *América Indígena* 30:653–670.
  1983   Comparison of Agrotechnologies in the Basin of Mexico and Central Maya Lowlands: Formative to the Classic Maya Collapse. In *Highland-Lowland Interaction in Mesoamerica: Interdisciplinary Approaches,* edited by Arthur G. Miller, pp. 13–47. Washington, D.C.: Dumbarton Oaks Research Library and Collection.
Tylor, Edward B.
  1861   *Anahuac: Or Mexico and the Mexicans, Ancient and Modern.* London: Longman, Green, Longman, and Roberts.
Vargo, Joe
  1986   Mexican Find Stirs Re-evaluation of Ancient Olmecs. *Austin American-Statesman,* 28 April.
Wagner, E. Logan
  1986   Teopantecuanitlán: A Newly Discovered Olmec Site in Western Mexico. Paper presented at the Institute of Latin American Studies Coyuntura/Conjuntura, The University of Texas at Austin, 6 November.
West, Robert C.
  1949   *The Mining Community in Northern New Spain: The Parral Mining District. Ibero-Americana* 30.
Whalen, Michael E.
  1981   *Excavations at Santo Domingo Tomaltepec: Evolution of a Formative Community in the Valley of Oaxaca, Mexico.* Ann Arbor: Memoirs of the Museum of Anthropology, University of Michigan, 12.

White, Lynn
    1966   Pumps and Pendula: Galileo and Technology. In *Galileo Reappraised,* edited by Carlo L. Golino, pp. 96–110. Berkeley and Los Angeles: University of California Press.

Whitecotton, Joseph W.
    1977   *The Zapotecs: Princes, Priests, and Peasants.* Norman: University of Oklahoma Press.

Whitmore, Thomas, and B. L. Turner II
    1987   Population Reconstruction of the Basin of Mexico: 1150 B.C. to present. Technical Paper No. 1. Millennial Longwaves of Human Occupance Project. Worcester: Clark University, Graduate School of Geography.

Wilken, Gene C.
    1969   Drained-Field Agriculture: An Intensive Farming System in Tlaxcala, Mexico. *Geographical Review* 59:215–241.
    1985   A Note on Buoyancy and Other Dubious Characteristics of the "Floating" Chinampas of Mexico. In *Prehistoric Intensive Agriculture in the Tropics,* edited by I. S. Farrington, pp. 31–48. Oxford: B.A.R. International Series 232.
    1987   *Good Farmers: Traditional Agricultural Resource Management in Mexico and Central America.* Berkeley and Los Angeles: University of California Press.

Wilkerson, S. Jeffrey K.
    1980   Man's Eighty Centuries in Veracruz. *National Geographic Magazine* 158:202–231.
    1983   So Green and Like a Garden: Intensive Agriculture in Ancient Veracruz. In *Drained Field Agriculture in Central and South America,* edited by J. P. Darch, pp. 55–90. Oxford: B.A.R. International Series 189.

Wilkinson, R. G.
    1973   *Poverty and Progress: An Ecological Model of Economic Development.* London: Methuen.

Willey, Gordon R.
    1981   Recent Researches and Perspectives in Mesoamerican Archaeology: An Introductory Commentary. In *Supplement to the Handbook of Middle American Indians,* vol. 1, *Archaeology,* edited by Victoria Reifler Bricker and Jeremy A. Sabloff, pp. 3–27. Austin: University of Texas Press.

Williams, Barbara J.
    1984   Mexican Pictorial Cadastral Registers: An Analysis of the Códice de Santa María Asunción and the Codex Vergara. In *Explorations in Ethnohistory,* edited by H. R. Harvey and Hanns J. Prem, pp. 103–125. Albuquerque: University of New Mexico Press.

Winkley, Brien R., Stanley A. Schumm, Khalid Mahmoud, Max S. Lamb, and Walter M. Linder

198   *Bibliography*

1984   New Developments in the Protection of Irrigation, Drainage, and Flood Control Structures on Rivers. In *Transactions of the Twelfth International Congress on Irrigation and Drainage*, pp. 69–111. New Delhi: International Commission on Irrigation and Drainage.

Winter, Marcus C.
1985   Los Altos de Oaxaca. In *Historia de la agricultura, época prehispánica, siglo XVI*, vol. 2, edited by Teresa Rojas Rabiela and William Sanders, pp. 77–125. Mexico City: Instituto Nacional de Antropología e Historia.

Withers, Arnold M.
1941   Excavations at Valshni Village, Papago Indian Reservation. M.A. thesis, Department of Anthropology, University of Arizona.
1963   A Survey of Check Dams and Archaeological Sites on the Albert Whetten Ranch, Chihuahua. Paper presented at the Annual Meeting of the Society for American Archaeology, Boulder, Colorado.

Wittfogel, Karl
1957   *Oriental Despotism: A Comparative Study of Total Power.* New Haven: Yale University Press.

Wolf, Eric
1959   *Sons of the Shaking Earth.* Chicago: University of Chicago Press.

Wolf, Eric R., and Angel Palerm
1955   Irrigation in the Old Acolhua Domain, Mexico. *Southwestern Journal of Anthropology* 11:265–281.

Woodbury, Richard B.
1960   The Hohokam Canals at Pueblo Grande, Arizona. *American Antiquity* 26:267–270.
1962   Systems of Irrigation and Water Control in Arid North America. In *Akten des 34 Internationalen Amerikanistenkongresses 1960*, pp. 301–305. Berlin: Verlag Ferdinand Berger, Horn-Wien.

Woodbury, Richard B., and James A. Neely
1972   Water Control Systems of the Tehuacan Valley. In *The Prehistory of the Tehuacan Valley*, vol. 4, *Chronology and Irrigation*, edited by Richard S. MacNeish and Frederick Johnson, pp. 81–153. Austin and London: University of Texas Press for the Robert S. Peabody Foundation.

Zimmerman, Josef D.
1966   *Irrigation.* New York: John Wiley and Sons.

Zipf, George Kingsley
1949   *Human Behavior and the Principle of Least Effort: An Introduction to Human Ecology.* Cambridge: Addison-Wesley Press.

Index